TO THE POWER OF ONE

Kabbalah Centre Publishing is a registered DBA of
Kabbalah Centre International, Inc.

For further information:

The Kabbalah Centre
1062 S. Robertson Blvd., Los Angeles, CA 90035
155 E. 48th St., New York, NY 10017

1.800.Kabbalah www.kabbalah.com

Printed in Canada, July 2016

ISBN: 978-1-57189-953-8

Design: HL Design (Hyun Min Lee) www.hldesignco.com

TO THE POWER OF

RAV BERG

KABBALAH
CENTRE
PUBLISHING

For my wife

Karen

in the vastness of cosmic space

and the infinity of lifetimes,

it is my bliss to share a soul mate

and an Age of Aquarius with you.

FOREWORD

When the Rav wrote this book in 1991, the world of medicine was not as it is today or as it will become.

The Rav took a hard line in the attempt to create a revolution of thought with regard to medicine, science, technology, and consciousness, so that we will be able to understand wholeness as opposed to fragmentation.

In an effort to edit this book, we, at Kabbalah Centre Publishing, took note of how much has actually changed and realized that much of that change is due to bodies of work like this one by the Rav and others, who fought the science and intelligence of the day. For the reader, we ask that you put yourself back in time when this book was written to fully understand not only the greatness of the Rav's thoughts and words but also the courage it took to stand out alone as a challenger of established thought.

History has shown us that what is scientific for one generation is later disproven upon the advent of new information. In fact, the science of a generation is more often a product of its culture than a product of science—the world is flat is one glaring example.

Those who challenge the convention of their day are the leaders whose questions initiate the righting of so many wrongs.

As you read this, please also remember that to overcome established paradigms means to take a strong position. Consider those change makers in history and the voice of their revolution. The Rav is no different.

It is a privilege to have the perspective of hindsight—to see where the Rav was guiding us, and how meaningful it still is. How, today, more than 20 years later, the wisdom found in this book still pushes us to expand our consciousness of life.

TABLE OF CONTENTS

Preface xiii
Introduction xvii

Chapter 1: Cosmic Attack 1

How the Cosmos manipulates man; karmic and astrological effects in our present lifetime; man's freedom to alter his destiny; our short-sighted approach to physical and mental well-being; harmony between the celestial and our terrestrial world; the composition of inner soul; the interplay between body-consciousness and the celestial and terrestrial bodies; vulnerability and security shields; the concept of empty space; the function of the five major senses; the necessary role of the individual in social, economic, and environmental affairs.

Chapter 2: The Mind-Body Connection 39

The promise of Isaiah; man as a determinator of physical and metaphysical activity; the application of Restriction to transcend the physical; comparison of the energy-intelligence of the body with that of the spirit; unseen influences; the power of thought; conscious control of cosmic influences and bodily states; prevention of vulnerability; mind and mind-brain; the effect of prior lifetimes on the present; the requirement to complete our *tikkun*.

Chapter 3: Human Myopia 79

The shortsightedness in dealing with hazardous waste and pollutants; the kabbalist's wholistic approach to the problem of human myopia; the mistake of splitting the atom; the reality level of order and eternity in Nature; the necessity for open-mindedness in the New Age; man's ability to correct the state of the environment and change to more positive behavior; the role of the *Zohar* in bringing mankind to a state of harmony.

Chapter 4: Fission or Fusion 121
The differences between the process of fission and fusion; the destructive elements of fragmentation resulting from the concept of fission; the concept of unity emerging in the new age of physics; the role of cosmic intelligence in the process of change; the negative activity of mankind bearing on the choice of fission; the holistic fusion approach described by the *Zohar*.

Chapter 5: Stress 143
The generally accepted understanding of the relationship of stress to physical and mental problems; defining stress; the responsibility of the individual to explore and alter the influences in the development of illness; the great transition phase now taking place; knowledge as the basis for change in the Aquarian Age; Light as the origin of stress; Restriction as the prerequisite for control of the Lightforce; the transformation of body-consciousness needed to relieve the crises of stress.

Illustration: The *Sefirot* 176

Chapter 6: The Outer and Inner World of Man 179
Soul- and body-consciousness; Man as the determining producer and director of movement in the universe; The Lightforce and its channels; the effects of being blind to the Lightforce; the role of the signs of the Zodiac and the effect of the body-consciousness on the Celestial region; the role of soul-consciousness extending beyond the Celestial region; human intelligence; the availability to humans

of the infinite intelligence of the Cosmos; the effect of prior
incarnations on the mind and consciousness; the mind and brain;
the function of sleep; the immune system and the function of the
thymus gland, the removal of veils (*klipot*) by Restriction; creating a
higher IQ; the goal to subject the conscious mind to the authority
of soul-consciousness; the healing process of soul consciousness; the
Encircling Light, the infinite reality; the importance of connecting
with and knowing the Life-force; the Central Column as the test of
connection to the Truth.

Chapter 7: Immune-Vulnerable 243
The metaphysical security shield; the importance of positive attitudes;
the creation of vulnerability by failure in the *tikkun* process, the effect
of Restriction on the functioning of the immune system; the
knowledge of cosmic time zones in avoiding vulnerability; going
beyond the external manifestation to the primary cause; taking
matters into our own hands.

Chapter 8: Time Travel 273
The possibilities of traveling at the speed of light; some flaws to be
expected in the experience of returning to the past; the kabbalistic
proposal of parallel universes; the illusionary universe and the Endless
reality level; the state of time, space, and motion at the Revelation
of the Force; the *Zoharic* solution to travel at speeds faster than light;
the Tree of Life consciousness.

References 304
Glossary 314

PREFACE

In 1971, the author of this book, Rav Berg, had been the Dean of the Research Centre of Kabbalah in Jerusalem for two years, when something took hold of his heart. He envisioned sharing the wisdom of Kabbalah with the world—a revolutionary idea. I say revolutionary because warnings about the dangers of entering this area of knowledge had been so dominant over the years that these warnings succeeded in keeping Kabbalah hidden. Perhaps a factor that contributed to its concealment was that world consciousness was not ready for it until now. And so, following his heart, the Rav began to set up Kabbalah Centres in parts of the world, attracting the assistance of caring souls with a strong desire to share this knowledge and wisdom.

At the same time that the Centres were being established throughout the world, the Rav's pen desired to write, and so in the stillness of the night it began. This is the fifteenth book on the subject of Kabbalah that the Rav edited or wrote. Kabbalah is a vast subject and there are many aspects to explore and many books still to follow.

There are two kabbalistic concepts that assist us to be in oneness with ourselves and our universe: One is the concept of Restriction, and the other is the concept of the Desire to Share, which completes the circuit starting with the Desire to Receive. These two concepts can be unified into one: "Love your neighbor as yourself," as Rabbi Hillel said, "Everything else is commentary."

Knowing the "why" of Creation helps us to avoid robotic behavior and to dislodge our misconceptions acquired often without thought. One of the prevalent misconceptions that damages our progress toward oneness and harmony is the notion of a Supernal Being that rewards and punishes. This concept is so ingrained within that it will take time and effort to realize that it is not true. Kabbalah teaches that all of the effects we experience are the result of our own thoughts and actions, even though planetary and environmental influences enter into the equation.

Wisdom about how to adjust our thoughts and actions to be harmonious with Creation is a thread running throughout this book, as well as all the other books written by the Rav.

The *Zohar* describes the reality of existence differently from what is generally accepted. Consequently, if you find yourself confused in the midst of a chapter, do not be discouraged. Keep going. Chances are your questions will be answered by the time you reach the end of the chapter. This was certainly my experience.

"Love your neighbor as yourself" comes from the Bible, Leviticus 19:18. This verse is followed by two additional words: *Ani Hashem*, which is translated as "I am the Lord." When these additional words are taken into consideration, the meaning of the verse changes to read: "When you love your neighbor as yourself, you will meet Me."

According to Kabbalah and Kabbalistic Astrology, we are well into the Age of Aquarius, with expectations of significant expansion in world consciousness. But for those who are not aware, the new age does not exist. The Rav's writings enliven our state of awareness so that all may participate fully in the Age of Aquarius.

Roy Tarlow, Editorial Staff
The Kabbalah Centre, New York, 1991

INTRODUCTION

The critical need to confront the world-wide problem of stress has become painfully evident by the enormous interest that business and government have taken toward finding methods to relieve and reduce this plague sweeping over our society. Stress has long been suspected of contributing to the number one killer of people, namely heart disease.

For all its action and glamour, today's business world and modern lifestyles generate conditions that wear down the body and spirit. Pressures are taking their toll by taxing the body's essential resources and draining its energy. Repeated exposure to the strains of stress brings on a process of heightened deterioration, which saps our very foundation. Marriages suffer, nervous breakdowns are commonplace, and careers are ruined.

As tension mounts and blood pressure rises, that extra drink or two at the end of the day becomes a habit. At a time when mergers, acquisitions and takeovers are "business as usual," no one can be sure when and where the dreaded takeover will strike. Today, this chronic, unremitting strain puts a cost to the economy, some say, as high as 150 billion dollars.

Some people are suffering from anxiety, mental illness, drug and alcohol abuse in a turbulent search for stimulation. Despite the physical/material comfort achieved through advancement, technology has, thus far, failed to enable man to achieve the personal satisfaction that he craves.

To the Power of One presents the kabbalistic perspective and understanding of the ills and traumas that beset our modern society. The viewpoint of the kabbalist is that the *individual* creates his own particular energy field of contentment or the lack of it.

The first step necessary to comprehend our dilemmas and misfortunes, from the quantum level, is to draw our attention away from the thought that "someone out there does not like me."

If indeed a negative environment exists for us, we have caused its expression and its influence over us. More importantly, just as we understand that the painful pressures and negative circumstances that plague mankind originate from within our own energy fields, so too, do the solutions to these problems emerge from our own *individual* efforts as well. The dismal track record of outside intervention testifies that invasion of our personal space often fosters disease. Leaning on a therapist without our own involvement perpetuates the problem.

To the Power of One shows us how we can regain our inalienable right of self-determination, and presents a vivid picture of how we can change our robotic behavior to one of conscious control and satisfaction.

Some social scientists propose a re-evaluation of the nature of human functioning. The mechanistic perspective suggests an engineering approach to health in which mental and physical illness are treated as mechanical problems.

The body's inherent healing ability, and its natural tendency to support a state of health, is not what is communicated by the medical profession, nor is the notion that we are self-healers. We assume that physicians can accomplish just about anything concerning our mental and physical well-being. Yet physicians themselves suffer from the mechanistic, fragmented view of health. Medical education and training generates stress without the education of how to cope with it.

When the mind is considered to be separate from the body, dis-ease is understood as a malfunction of our physical body, in no way connected with an integral aspect of our thought process. And yet the word "dis-ease" itself is a definitive clue to ill health, namely that stress and dis-ease are one and the same. Webster's definition of *ease* is: "freedom from pain, worry, trouble or strain." The prefix *dis* means "the opposite of." Consequently, the medical profession termed physical ill-health to be a condition lacking "freedom from pain, worry, trouble or strain."

Because of its difficulties in dealing with the psychosomatic aspect of medicine, the medical profession cannot bring itself to face the reality of its influence. This lack of correlation and connection between the mind and chronic degenerative diseases, such as cancer and heart disease[1], results in physicians accepting that these incurable afflictions are the inevitable consequences of general deterioration. They refuse to expand their frame of reference.

The current medical approach, which fragments mind and body, is grounded in Cartesian thought.[2] René Descartes

(1596-1650), seventeenth century mathematician and philosopher, introduced the concept of the absolute separation of body from mind; the body was to be considered and understood completely in terms of an arrangement and function of its parts, no different from a well-made complex machine.

A person was considered healthy if all his parts were operational, and sick when one or more of his parts were not functioning properly. And since that time, medical research has limited itself to understanding the physical mechanisms involved in the disease of any part of the body. This approach has left out all the possible influences of a metaphysical, non-corporeal nature on the biological process. The physician's role is to invade, either surgically or chemically, the malfunctioning part and correct the specific mechanism.

Three centuries after Descartes, the attitude that the body is nothing more than a machine, that disease is a consequence of the breakdown of this machine, and that medical-biological research can fix it still dominates the world of medicine.

The misconception that governs the sciences of the physical reality is that the human body can be treated as an entity of an inanimate nature, nothing more than a machine. The logical consequence of this perception of the body is a medical system that can at times be insensitive and even degrading.

From a kabbalistic point of view, we are 99 percent metaphysical thought-consciousness, which means that the physical matter of our body structure occupies only one percent of our space. Therefore, when our medical system concentrates on smaller and smaller fragments of this one percent, namely the body, it is inevitable that we will lose sight of the patient as a human being. It is not a wonder then, as we reduce diagnosis and health care to such a diminutive understanding of man that such little progress has been made in our overall general health. How can we possibly neglect the metaphysical and mind components of the human being when they comprise the 99 percent of who we are? Although scientific research reveals that attitude is an essential aspects of all medical treatment, the mind remains outside the scientific framework of medicine.

The reason the mind is rejected is simple and evident. Since western scientists have no idea what or where the mind is, they cannot address themselves to it. What cannot be weighed or measured, put into a test tube or scanned with a mass spectrometer does not exist. This perspective brings with it serious implications.

To reincorporate the idea of mind healing into the theory and practice of medicine, medical science will have to transform its narrow view of health and illness. Does this necessarily mean that science will have to settle for anything less scientific? The answer is, of course, no. However, science will have to broaden its conceptual base to include phenomena that are correct despite the fact that they cannot currently be scientifically verified.

The history of modern medical science demonstrates that the reduction of human life to test tubing and molecular observable fact has influenced the practice of medicine. It split the profession into two separate and distinct groups: Physicians who are concerned with the treatment of the material body, and psychiatrists, physicians who deal with treating the mind.

To me, this represents the reason that medical research does not examine the role of stress in development of illness. The study of stress is generally confined to the psychological aspects of illness, despite the overwhelming evidence that links stress to a wide range of diseases and disorders. For example, cancer does not begin with the observation of a lump or tumor. Its beginnings are found in the past.

To aggravate the matter further, even with regard to our mental health and well-being, today's psychiatrists treat mental illness by physical means. They attempt to understand mental illness in terms of physical, chemical problems in the brain. Unfortunately, the patient population in mental health facilities has not dwindled but rather increased.

The understanding of the effect of cosmic influences on our mental and physical well-being is long in coming. Nothing, as yet, regarding this subject has even been intimated by researchers. Mental illness is considered the same as other physical illness, only in the case of mental illness, the brain is affected rather than some other part of the body. Furthermore, mental illness manifests itself in unusual social behavior, and physical illness is contained within some part or organ of the body.

We are all familiar with cases where the patient is in a constant state of fatigue, just not feeling "good." Yet, after extensive testing, the patient was given a clean bill of health. Medically he was healthy, though he felt terrible. Because medical science is grounded in physiology, it rarely deals with the psychological aspects of sickness. The pathways of communication between mind and body are not fully understood, and even ignored.

Medicine has lagged far behind the new findings in physics. Physicists now tell us that in addition to the Cartesian paradigm—the foundation for modern medical science—there exists another level of existence, namely, that which cannot be subject to scientific verification: Our thought consciousness.[3]

The kabbalist has long known that *thought controls all manifested states of the physical reality*. In fact, the energy-intelligence as adapted to the corporeal, material world is illusionary. This has long been known to the physicist. Why all mankind has not been educated to this startling discovery truly remains a deep mystery. Just imagine the state of oneness that could exist in a society where this knowledge is deeply ingrained.

The concept of mind over matter is dealt with lightly in contemporary thought. However, the *Zohar*[4] gives expression to an important text in the Biblical Code pertaining to astral influences. When Abram (later named Abraham), the first astrologer, gazed up at the stars, he foresaw that he would not have children with Sarai (later named Sarah). The Lord told

Abram not to gaze any longer into the stars for he could have a son if he attached himself to the Upper Realm of consciousness and not to the energy-intelligence of the material stars.

As with everything in the coded text of the Bible, the central message is packaged within abstruse wording. This sometimes leaves us with the impression that the Bible merely consists of a new religion for the Israelites and a collection of ethical rules of conduct.

Kabbalah teaches us that we must be ever wary of appearances, for things in this physical world are never what they seem. Now, as always, the physical universe gives every impression of being in a state of perpetual darkness and chaos. The Light— the cure—is here, but is so obscured by the negative, material trappings of finite existence that it requires a sensitive eye and a compassionate soul to perceive it.

Even today, the astute observer can detect trends within western science that seem to indicate a swing away from the corporeal, material illusion. Where previously a cable carried 400 conversations, now a fiber-optic strand, no wider than a human hair, can transfer 400,000. From a kabbalistic perspective, the fundamental importance of this new technology is that it provides a conceptual framework, a jumping off point, if you will, for mankind to place less emphasis on the material world and to connect with the real world—thought energy-intelligence. But what about the reality of pain, cancer, and other terminal diseases? Are they not real? How are we to accept these very material concepts within a frame of illusion? How do we go about informing a loved one that little time is left before death? How can we minimize or even remove the scourge of pain and suffering that has become a common occurrence within our familial scene of reality?

To reassure you, these questions and their significance are dealt with in this book. Remember the old saying, "It is all in the mind." Metaphysical connections are invisible by their very nature, so much so, that most of us are ignorant of them. However, those brief encounters with other dimensions provide evidence of a world so superior to this phase of existence that comparison is hardly possible. Having no words to describe our experiences, and no validation by our present culture of the existence of higher realms, we banish the memories of our extraterrestrial sojourns to hidden catacombs deep in our subconscious mind.

The reason I have devoted so many pages to this perspective of illusion and reality is to bring home the notion that, for the most part, scientist and layman alike operate via robotic consciousness. Furthermore, we have treated the essential nature of material reality as a fragmented structure. This view has led all of us to be drawn to what I refer to as the "quantum of symptom disorders." We tend to treat the superficial, outward symptoms of illness, whether they be physical, mental or sociological, without paying enough attention to the internal, metaphysical causes of these external disorders. Temporary relief has become the by-product of this misleading, corrupted perspective of the life sciences.

Fortunately the information revolution brought on by the Age of Aquarius is working in our favor. A major sector of Earth's inhabitants no longer prefers the accepted treatment of medication that controls the symptoms of disorder but does not cure it. Removing symptoms by repressive measures has left our society in a state of dilemma. To associate a particular ailment with a definite part of the body draws our attention away from the patient as a whole person. Also, to associate a particular societal problem with its components draws our attention away from society as being a part of the universal whole—and this totality must take into consideration the effects of the influence of the cosmos.

Although conventional models do distinguish between symptoms and diseases, each disease (or problem in a broader sense) should be seen as a symptom of the underlying cause, whose origin is rarely explored. For this reason, holistic medicine is now becoming respectable. Now that neurological and chemical connections have been discovered between the brain, pain, and the important immune system, greater emphasis is being placed on a more holistic approach to the future welfare of patients, society, and the environment.

This book deals with the power of Kabbalah and its relation to our daily lives, a subject that has not been addressed before. Although the wisdom and its benefits have long been known to the kabbalists, the knowledge contained in the Kabbalah has been withheld from the general public. Some have mistakenly labelled its model as belonging to the Israelites or the religion of Judaism. Others have guarded its secrecy, considering its teachings too Divine or holy for just anyone to enter.

Medicine, biology, psychology, and the environment play a decisive role in all universal models established by the life sciences. However, principles determined by research experience must, of necessity, include the effect of astral influences. To ignore their influences and the important role cosmic energy-intelligences play upon the universal stage of humanity is to invite helplessness and hopelessness into the daily lives of mankind.

Among the general public today, there is widespread dissatisfaction with doctors, psychologists, and government. Most people are not aware that one of the main reasons for their current state of affairs is the narrow base from which solutions spring forth. The overwhelming majority of illnesses, as well as sociological and environmental problems, cannot be understood or defined in terms of the single cause. The main error, from a kabbalistic perspective, is the approach our leaders are taking to these problems. Instead of asking the most important and singularly significant question, namely, "why" these diseases or problems occur in the first place, and then making an attempt to remove these conditions, the focus is directed toward the manifested mechanisms through which the problem or disease operates, in order to alter or interfere with them.

This confusion and its approach are at the very center of all universal problems. To go beyond this way of thinking, and achieve the kabbalistic model for problem solving, will require nothing less than a profound educational revolution. However, most people are afraid of change because this means a re-examination of their lifestyles. Very few people are prepared to change their ways despite the inevitable track record that history furnishes.

What people do not understand is that it is not *lifestyles* that the kabbalists address. Lifestyle is essentially not threatened. What the kabbalist strives for is a change in human behavior. There are but two alternatives: To continue with the same universal model that has always brought chaos, disorder, and disease to our lives or accept the cosmos as an integral force in our lives through which all this can change.

This will require public re-education. In the Age of Aquarius, there are very few choices available. Transcending the universal model will be possible, claims the Kabbalah, only if a holistic approach is incorporated within a system. This system must address itself to the individual so that he or she can achieve a purer state of awareness and an altered state of consciousness, where we again gain control of our lives and destiny, including, as well, control over the cosmos.

By attempting to make the study of Kabbalah available to everyone throughout the world, The Kabbalah Centre hopes not only to assist individuals in achieving altered states of consciousness but also to assist in creating a personal security-shield that can protect from a hostile environment, should society continue its present trajectory of insanity and chaos. Specific objectives of study include the development and expansion of the frame of reference. In this way, we increase our awareness, improve our chances for success in new undertakings, avoid the pitfalls that bring on crisis in family life, and above all, avoid ever having to resort to such expressions as "lucky" and "unlucky."

If we think about it, for the most part, we feel at a loss when it comes to controlling crime that strikes our neighborhoods. A criminal will strike one home but not all potential targets. Why is one home chosen instead of another? The accepted response is usually that the victim of this specific house was unlucky and the others lucky.

One car on a highway is involved in an accident, while all the others travel along without incident. Why did the drunken driver appear at that moment, not a second sooner or later? Again, the answer given makes reference to the "lucky" or "unlucky" ones.

The study of Kabbalah has emerged at a time when the fear of AIDS, cancer, and other maladies are in the minds of mankind.

What makes one person vulnerable while another remains unaffected under the same external circumstances? How do our thoughts, attitudes and feelings affect our health? Cancer, for example, is not an invader from the outside that strikes the body. Cancer manifests as a cell behaving in a distorted manner within our bodies. Why does one person experience malfunctioning cells? Why, in one individual, does the immune system recognize abnormal cells and destroy them or at least confine them so they cannot spread, whereas in another person, the immune system is not strong enough and the cell reproduces others with the same imperfection?

Cancer is not an enemy from without but a negative force from within. There is no doubt that carcinogenic substances may contribute to the formation of cancer cells. However, we must come to the realization that neither carcinogens nor other environmental influences alone provide an adequate answer to the causes of cancer. After all, we are all exposed to these same conditions.

No answer is complete without addressing the question: Why one and not the other? Are we to rely on the familiar explanation of lucky or unlucky? To only consider the mental and emotional aspects of health and illness, without examining the influences of the cosmos and human behavior is, from a kabbalistic world view, an exercise in futility.

There is a great deal of confusion surrounding the myth of stress-related symptoms and causes. Primary influences are the subject of much research, and I might add, there are as many primary influences as there are research projects. To address ourselves to just a handful of them requires an enormous amount of time and effort. And then, who knows whether we have been directed to the right ones?

Unhealthy diets, the cumulative effects of chemicals, preservatives, and additives that we ingest or inhale are all seen by many as primary influences affecting people. There are also unseen influences that can be considered as primary or secondary. However, for every documented case of an unseen yet real influencer, there is always the living demonstration that the influencer does not affect health as we might expect.

Let us also remember that modern living, for all its above-mentioned failings, provides a life expectancy greater than at any time during the past three hundred years: from a low of 35 or 40 years at the end of the 18th century to the present life expectancy of 72 years. These facts present an apparent contradiction.

At this point, the question previously raised, which cannot escape our attention, is whether some or all of us are victims of circumstance. Are we merely puppets whose strings are manipulated one way or another? In what sense are we in control of our lives and to what degree are external forces affecting our life expectancy and our state of well-being?

From a kabbalistic point of view, the answer to control lies in our relationship with, and our awareness of, the cosmos, which is the primary influence in our lives. Our attitude, and the state of our mental and physical health, most certainly affect the outcome of our lives and the way we live it; however, they are not the primary source of the energies that evolve and become part of our lifestyle. Everyone has a desire to be positive and to love, at least, those closest to us. And yet, we are not in control of our actions as indicated by the alarming cases of child neglect and abuse, and the maltreatment between husband and wife and within families.

During a period of negative astral influences even appropriate actions can have negative results, and conversely during a period of friendly cosmic influences even illogical actions can turn out successfully. *The cosmos is primary, everything else is secondary.* Let us examine some of the kabbalistic references substantiating this new and revolutionary approach to the world's ills and tragedies. The *Zohar* states:[5]

r every man who is compounded of the four elements [fire, air, water and earth] is accompanied by four angels [positive extraterrestrial energy-intelligences] on his right side and four on his left.

One of the four angels [on the right hand] will be Michael... one Gabriel... one Nuriel... and one Raphael.

The four powers of evil [negative extra-terrestrial energy intelligences on his left]... will be Anger, Destruction, Depravity [wrong doing], and Wrath.

... and from the aspect of the body [manifested state] the Angel Metatron presses close to him from the right side and Samael [Dark Lord] hovers above at the left side.

Now all men are formed of the four elements, but on the order of the constellations with which each man is connected depends the order of the angels who accompany him, and also the potential characteristics of the man. Thus, if his ruling constellation be the Lion [Right Column or water sign] Michael will be dominant, and be followed by Gabriel, and after him Raphael, and lastly Nuriel. If, however, his planet is the Ox [Taurus, fire, Left Column] first comes Gabriel, then Michael, then Nuriel, and then Raphael. If the Eagle [air sign be his constellation by which he is influenced, Nuriel will be first [in domination], then Michael, followed first by

Gabriel and then by Raphael. And should his constellation be Man [earth sign] then Raphael will be dominant, with Michael, Gabriel and Nuriel coming after in the order named.

The *Zohar* continues to detail the precise composite of an individual and his or her behavior, depending on the birth sign. Our life existence has already been cast based upon our prior incarnations. Reincarnation is the mold by which our lives are sculptured. Its manifestation takes place by a predetermined cassette.

Kabbalah is a science and study that permits the individual to control his destiny, plug into the major regenerating powers of the subconscious and avoid unchecked stress and serious illness. There are no mysteries. An exact description of each human being is further elaborated upon in the *Zohar* just mentioned.

The point that is stressed in the *Zohar* is that we are inextricably bound and linked to the cosmos. Our behavior depends completely on our constellation sign—our strong qualities are noted, as is the methodology by which our weak characteristics may be strengthened. *Navigating the Universe,* previously know as *Time Zones,* provides a timetable of cosmic danger zones and friendly skies. This knowledge affords us the opportunity to tap the awesome power of the cosmos, restructure our life cassettes, and splice where splicing is necessary to avoid the secondary influences that wreak havoc upon our civilization and create chaos and disorder in our lives.

Chapter One

COSMIC ATTACK

Chapter One

COSMIC ATTACK

WE ARE UNDER CONSTANT ATTACK FROM THE COSMOS. Consciously, or otherwise, we are at its mercy. Its constant bombardment of infinite thoughts is relentless: These thoughts disturb us when we want to go to sleep; when relaxing after a difficult day, we are not permitted the luxury of a quiet mind despite all our efforts at quieting the mind. To push aside, even for only a few moments, the thoughts the cosmos imposes upon us becomes a monumental task. After a while, we just resign ourselves to it. The battle is too furious; there is no sign of relief. For the most part, and for most of us, we find ourselves involved in an exercise in futility. There seems to be no way out other than escaping to some other activity.

Only when we realize the extent that we are subjected to external manipulation, can we begin to improve our physical and mental well-being. Does the cosmos influence and affect all people in the same way? If not, why not? The cosmic ensemble of celestial players are maneuvered by the all-embracing reality, the Lightforce. Their rotations, movements, and revolutions appear on a regular predetermined schedule, and we are all influenced by their astral intelligences in different ways.

What determines how mankind is touched by these extra-terrestrial influences in unique, individual, and particular ways? What factor is employed that almost forces us to behave in a manner that causes us to turn back and ask, "Why did I do that?"

The kabbalist raises the question, "What caused what in the first place?" despite the *Zohar*'s conclusion that our universe and the lives of all mankind are programmed by cosmic computerized printouts. Nevertheless, how do these printouts take on the individual characteristics that program each particular aspect of our universe in their own particular way?

Reincarnation, the cassette of prior lifetimes, is responsible for a person's creativity, free-will, and emotions, such as love, hate, fear, and warring instincts. This cassette has an intelligence of its own—based on past behavior, the positive and negative actions create a new metaphysical DNA which is the embodiment of all these prior actions. The interface between the physical and metaphysical realm, between the present and the past, is the lineup of astral bodies at the time of one's birth.

The kabbalistic perspective of astrology is dramatically different from the conventional pursuit of this science. Conventional astrology contends that the individual will take a course of action *because* of the arrangement of the stars. Kabbalah contends that the *tikkun* process places the individual in an astrological position so that the stars will impel the individual in the necessary direction.

Birth charts are a pictorial view of this metaphysical interface. The physical celestial entities do not determine or affect the metaphysical realm. It is always the unknowable, non-material realm that ultimately manifests as these particular channels of energy by which our physical, material life existence becomes a reality.

These channels of energy provide the varied scenes of our past lifetimes. When we have behaved in a negative manner, during our present lifetime at a particular moment corresponding to our prior incarnation, we will become infused with negative energy-intelligence. Our prior negative activity is being superimposed on our present life experiences to provide us with another opportunity to make a correction, a *tikkun*.

Each day of a present lifetime will correspond to the exact time of our former lifetime. If, for example, one's birthday at age 26 is today, the astral bodies will transmit the identical printout of the behavior on this date in a prior incarnation. The interface deals with combinations of complex Life-forces that subsequently become manifest in our present physical realm.

At the time of birth, the positioning of the complex mechanism of planetary bodies acts as a physical interface that encompasses the complete printout of prior lifetimes. This process may be compared to what takes place at the time of conception when the sperm of the male and the egg of the female contribute their own printout to the developing embryo.

When we read in the *Zohar* that these astral influences have a profound effect on us, it is because the reincarnation process

5

is based upon the character flaws of a former lifetime. The exact time and place of birth reveal the individual's primary, and I stress primary, life pattern, such as potential power, attachements, and problems.

For all the inviolability of the basic destiny pattern, we have a degree of freedom that is almost without limitation. We can determine how the *tikkun* process will take place within our present lifetime. The natal chart reveals the blinders and restrictions that will prevent us from feeling free. In addition, shortsightedness, bigotry, and former non-spiritual attitudes may prevent us from making use of the tools that are available by which we can transcend to another level of consciousness and make a change in our destiny.

These blinders are of our own creation. Opposition to Kabbalah by some today is merely an involuntary position taken in some prior lifetime, and expressing itself again. We built these barriers before. However, because we manufactured them, we can shatter them and ascend to higher levels of consciousness.

Let us return for a moment to the cosmic bombardment that most of us experience during times of relaxation. What is the source of these complex and diversified thoughts that appear in our subconscious? Inasmuch as the subconscious does not reason but acts according to the input programmed from a prior lifetime or many prior lifetimes, these thoughts are an accumulation of what was going on then. In addition, the innumerable impulses that weave their way through the fabric of the mind become an integral part of the thought process.

Comparing this process with that of a computer, we find that with a computer the machine does the thinking for us, whereas with our mind we are the process itself. Impulses and electronic stimulations enable a functional computer to achieve the final output and printout. Our subconscious mind operates on a similar principle, with the exception that we humans have the ability to become aware of the stimulations as expressions of cosmic attack.

The same process occurs in our physical bodies on a terrestrial level. Physical or emotional stimulation of our bodies or psyches elicits different responses, some subtle, some more apparent. In the metaphysical universe, stars neither shine constantly nor transfer energy without cessation. Rather they are radiant only at appointed intervals.

The *Zohar* states, "Each unit of consciousness or intelligence returns to its former unmanifested position after having served its purpose."[6] Thus, both our mundane universe and our own physical bodies reflect a constant back and forth movement between terrestrial reality and the reality of celestial systems. The cosmic realm constitutes the timeless, spaceless realm we must reach if we are to be true masters of our own destiny.

Once again, the *Zoharic* truth, "As above, so below,"[7] is demonstrated: We may paraphrase it to say, "As in the metaphysical universe, so in the physical."

To illustrate the point further, let us reflect for a moment on the response of an atom to outside stimulation. Some of the electrons, when stimulated, become excited and respond by moving into a higher orbit, farther away from the nucleus. Remove the stimulation and they will drop back into their former orbital cycles. Penetration and observation of deeper, more microscopic areas will ultimately let us see the rapid back and forth movements that take place in this process.

The barrage of thought energy-intelligence that we are subjected to originates in this lifetime from the cosmos. However, the Kabbalah reveals that even the cosmos is not considered the primary source for good and evil, health and disease. Stress and tranquility point up the need for source determination. So long as we continue to put our attention on the symptoms, we will fail to discover the underlying cause. The invasion of our bodies with medication and surgery to correct the symptoms of disease is consuming the creative energy needed to find the basic cause and take remedial action.

We have been programmed by our culture to believe that anything more than temporary relief is probably impossible. Thus we delegate all responsibility for our health to the *authorities*. There is an old familiar kabbalistic saying, "You get what you ask for."

The question that may be raised is as follows: "If indeed it is that simple, then why all the chaos and disorder? If the Lightforce is concerned about our well-being, why do we suffer? Why are we experiencing an ever increasing rate of heart disease, incidence of cancer and other degenerative diseases— all this despite the scientific advances of the past century?"

Transcending the basic model of society will be possible only if we are willing to change other things as well. An entire social and cultural transformation is necessary. Where and how is this educational revolution to come about? Who is going to be responsible for its momentum and successful conclusion?

The inhabitants themselves of planet Earth are the only participants in what I refer to as a necessary people's revolution. In the Age of Aquarius, we can no longer depend on government intervention or controlling bodies of authorities to extricate us from our present condition and various maladies.

It is the individual who can extricate us from the mess in which we now find ourselves. The failure of governments to stem the tide of human self-destruction by drugs is quite apparent. The problem of drug abuse is too monumental a task for government alone. The financial burden and the drain on human resources is too horrendous for any government or governments to undertake.

The present framework in the social sciences is inadequate. In light of recent developments regarding the plague of AIDS, our medical approach has become increasingly unrealistic. The scourge of worldwide diseases continues unabated despite the enormous accumulation of new information. Thus, while there has been progress in unraveling biological factors involved in specific diseases and in developing technologies that will affect them, identifying and labeling the disease does not necessarily mean making progress in health care.

This is not to say that in areas of emergency medicine, there has not been remarkable progress in dealing with acute infections, premature births, and the like. The spectacular medical procedures relating to organ transplants and open heart surgery do not answer for us the question as to why these conditions arose in the first place nor what measures may have *prevented* these physical impairments from occurring.

During the past hundred years, contemporary medical research has taken credit for the sharp decline in infectious diseases, such as cholera, polio, and tuberculosis. Today, most of these maladies have almost ceased to exist. However, Thomas McKeown,[8] a leading specialist in the field of public health and social medicine, provides sufficient proof that the striking decline in mortality was not the result of medical intervention alone. There were other contributing factors as well, including the improvement in hygiene, sanitation, and nutrition.

His study showed that major infectious diseases had all peaked and begun to decline long before the first effective combatant medications and immunization techniques were introduced. Many other research experiments seem to indicate that medical intervention alone is incapable of bringing about significant changes in basic disease patterns.[9]

This book is not intended as a condemnation of medical practices, its research or the physicians whose ideals brought them to caring for others and improving their physical well-being. The responsibility we have for the well-being of every individual was expressed in the universal code of the Bible: "If men strive together and one smites another... he shall pay for the loss of his time, and shall cause him to be thoroughly healed."[10] However, it appears from various commentaries on the Bible that the understanding uppermost in the mind of the physician must be that he is *only* a channel for healing purposes. The ability to heal depends on the level of spirituality of the physician and whether or not he practices his profession with a profound respect for human dignity and suffering.

Why does the Bible place such importance upon a physician's human spirituality? Why is the sense of consciousness stressed as a physician's prerequisite? In the view of the kabbalist, the physician's ability must, of necessity, be considered. However, inasmuch as his expertise is usually limited to the existing level of human potential (as low as five percent and as high as twelve percent), the kabbalist requires of the physician additional qualifications to increase his ability to heal others.

The kabbalist draws on the unlimited fields of energy, the enormous well-springs of information available with the expansion of awareness to cosmic consciousness. Cosmic consciousness is everyone's Divine birthright. The kabbalist explains this universality in terms of an intrinsic ability of the human mind and psyche to come in contact with cosmic consciousness. The term "cosmic" indicates and includes the necessary energy-intelligence information available to an individual at any given time.

The human nervous system and the brain comprise a highly complex system that remains deeply mysterious in many of its aspects, despite several decades of intensive research. The human mind is a living system *par excellence.* In the following chapter, we will discuss more fully its complexity and relationship to the body. For the present, the mind includes the ability to translate information of the remote past, and concern for the distant future.

Because of this unusual characteristic that distinguishes the human being from other animals,[11] we are innately and unconsciously drawn to the cosmos. Why? Because the cosmos contains the necessary ingredients for human fulfillment. As an extension of the Lightforce, the cosmos is primarily concerned with sharing its benefice, more than we have a desire to receive it. "More so than the new born calf wants to suckle, the cow desires to nurse."[12] The cow has a greater Desire to Share.

However, intermingled with its intrinsic Desire to Share, the cosmos also contains its portion of negative energy-intelligence. As an interface, the cosmos communicates our past lives with a full complement of both negative and positive activity. In our interactions with our environment, there is a continuous interplay and mutual influence between the celestial cosmic world and our terrestrial inner world of reality.

The cosmic pattern perceived by the kabbalist conforms in a very fundamental way to the patterns on our terrestrial plane. Man consists of two realities, the inner soul reality and the outer corporeal reality we know as our bodies. As human beings, we shape our environment very effectively depending on which of the two realities dominate our behavior. Our soul-consciousness possesses an energy-intelligence similar to that of the Lightforce, one of a Desire to Receive for the Sake of Sharing.[13] The soul is a metaphysical force that creates life within us. When the soul leaves the body, it creates death since there is no life in the body itself. Physical existence ceases to have purpose.

Our body-consciousness is a motivating channel for the Desire to Receive for Oneself Alone. The body is a physical entity, however, and there is something beyond the operation of cells and genetic makeup that makes it grow and function. This force is called body energy and, as the Desire to Receive for Oneself Alone, is the root of all evil because this force consciousness subjects individuals to the limitation of time, space, and motion.[14]

Its energy is the same as the energy of the Earth which, with the grip of its gravity, desires to swallow everything in its reach. Only when soul-consciousness has dominion over body-consciousness, does the body become integrated into the whole, converting the whole to a Desire to Receive for the Sake of Sharing.[15]

Each celestial body consists of an identical formation of inner soul and body-consciousness. Depending on our ability to tap cosmic positive energy-intelligence, we must also become aware of the negative energy-intelligence influence of the cosmos. The cosmos channels any available negativity to mankind. The methods by which we can avoid its attacks upon us are paramount to assuring our physical and mental well-being.

Another important consideration for our investigation into the causes of personal and universal chaos is vulnerability. In cases of serious accidents or cancer development, the kabbalistic world viewpoint is that vulnerability is the agent that is directly responsible for inviting and manifesting a negative cosmic attack. Let us again take cancer as an example. Everybody produces abnormal cells in the body from time to time. This can be attributed to either external factors or a distortion of cellular functioning.

Normally, the body's immune system, including the DNA, keeps close watch for any abnormal cells and destroys them. For cancer to occur, therefore, the immune system must be inhibited in some way. The important point here is that something is happening in the person who contracts the cancer to create a vulnerability. All of the factors mentioned previously, such as stress, the food we eat and the environment, may play a role in the incidence of cancer. However, none of them provides a full explanation as to why particular individuals, at particular points in their lives, attract the production of cancerous cells and, having attracted it, why their body's defenses are not functioning.

These same individuals have certainly been exposed to these same factors at other times. If there had been a genetic predisposition, it was there from day one. Any genetic condition of abnormal cell functioning is present throughout life. The crucial question is: "What happened to the body's defenses that permitted these cells to produce life threatening cancer cells at this moment in time? What prevents the immune system from performing a function that it had so successfully accomplished for so many years?"

There is no question about the strong links between the factors already mentioned and cancer. There is also no question that in our study of Kabbalah, we must consider the interrelationship of mind, body, and stress, and its influence upon illness in general. The question that must be raised at this time for all of mankind is "why me?" and "why now?" I am sure there are no cancer specialists who have not asked of themselves, "Why does one patient die and another, with the same prognosis and treatment, recover?"

The questions concerning illness and disease are, essentially, the same ones we might raise regarding automobile or airplane accidents or other misfortunes that might overtake us: "Why me?" and "why now?" However, before we continue to explore this most vital link into the causes of mankind's mishaps and tragedies, let us examine the *Zoharic* perspective on vulnerability.

> *A person should take care not to make himself visible to the negative, destroying forces when they swoop down upon the world, not to attract their notice, since they are authorized to destroy whatsoever comes within their view. This agrees with a remark of Rabbi Shimon Bar Yochai, that if a person is possessed of an evil eye, he carries with him the eye of the destroying negative force; hence it is called "destroyer of the world," and people should be on their guard against him and not come near to him. By avoiding him, one should not be injured by him.*

It is forbidden to come near him [with an evil eye] in the open [without any shield of protection]. If it is necessary to beware of people with an evil eye who can act as a channel for negative energy, how much more should one beware of the channel of death.

An example of a man with an evil eye was Bilaam, of whom it is written, "The speech of man whose eyes are open."[16] This means he was a channel of an evil eye and on whatsoever he fixed his gaze he drew the destroying negative force. Knowing this, he sought to fix his sight on the nation of Israel, so that he might destroy everything upon which his gaze should fall.

Thus it is written, "And Balaam lifted up his eyes"[17] so that his evil eye should fall on Israel. Israel, however, was immune; for it is written, "and he saw Israel dwelling tribe by tribe," (Numbers 24:2) and he also saw the Shechinah [protective shield] hovering over them. Because She made whole by the twelve tribes beneath Her, the eye of Bilaam could not have power over them.

He said, "How can I prevail against them, seeing that the Lightforce from on high rests on them and shields them with Her wings?"[18]

What seems to emerge from the *Zohar* is the startling revelation concerning vulnerability and the existence of security shields. Bilaam anticipated and searched for the vulnerable point towards which he could direct his cosmic attack upon the nation of Israel. His efforts, however, were of no avail. The people themselves controlled the negative energy-intelligence transmitted by the cosmos. Bilaam's ability to channel this enormous power of devastation was caught dead in its tracks.

The nation of Israel, as well as all of mankind, came into cosmic control at the time of the exodus from Egypt. Without this control over the Celestial Realm, persons like Bilaam could peddle their wares to unsuspecting, vulnerable victims without their ever being conscious of what was happening to them. The nation of Israel made good use of their cosmic teachings. Consequently, they were not open to the destructive forces by which Bilaam could devastate his victims. They were not vulnerable because they had placed their security shield in a position by which they could stave off any attack.

The assault may have originated with individuals capable of channeling negative cosmic forces or, if vulnerable, by the individual's personal negative activity of a prior lifetime. People who fail to activate their security shield become the victims of cosmic negative astral influences.

> When the letter appears to perfection, namely the Vav, which is the Central Column, as mentioned, all the Evil Sides are blocked and depart from the moon, which is Malchut, and do not cover it.[19]

17

What seems to emerge from the *Zohar* is that when we raise the question, "why me?" or "why now?" the answer lies at the doorstep of the victim.

The *Zohar* declares that ignorance of this primary cause, namely a lack of security shield, is at the heart of all mishap and disease. The critical need to confront the problem of cosmic channeling and cosmic influences cannot be ignored or underestimated. With proper and correct knowledge of the Kabbalah and of cosmology, we can move above the cosmic influence of the stars.

Within this context lies the answer to the crucial questions we asked before. For most of us, a concerted effort and concentration, as well as focused attention on cosmic danger zones,[20] provide the first safety and security measures to be taken to avoid the catastrophic misfortunes that presently befall mankind. When the *tikkun* process[21] dictates retribution and payment for past negative activity, at that moment in time, the individual must be prepared for an onslaught of his own negative influences. If, at that very moment, security shields are not placed in position, then we are vulnerable to the successive factors of stress, disease, and misfortune.

We have no one to blame but ourselves or those people who go around convincing the unfortunate that the study of Kabbalah "may be dangerous to our health." The "untimely" attack of disease upon our immune system is not as untimely as some physicians or scientists would have us believe.

Our present lifetime is merely a reproduction, in cassette form, of our previous incarnations. At the precise time when retribution makes its appearance and demand, we had better be prepared to correct our negative activity of a former lifetime. How are we to know when retribution is rearing its ugly head? Whenever our consciousness comes upon an opportunity for us to commit an immoral act, an act closely connected to a Desire to Receive for Oneself Alone.

If, at that time, we should fail in the *tikkun* process this time around, we have created for ourselves an empty space where the Lightforce has no connection. These spots are prime targets for the Darkforce and its armada of negative energy-intelligences that I call the death star fleet. In moments such as these, we have become vulnerable to an invasion,[22] and there is no telling how their barrage will manifest in our personal lives.

If and when the *tikkun* process opportunity presents itself, and we take the decided advantage, then, in essence, we splice out that portion of the cassette, and join the two ends connected with the Lightforce. The space that our negative activity occupies is, and relates to, the illusionary reality level of our existence where the aspect of free-will exists.[23] As long as this space-time frame of negative energy-intelligence has not been removed, there is injected what is known in kabbalistic terminology as empty space. The framework exists because the Lightforce has not filled this empty space with beneficence. Once beneficence or fulfillment is missing, there is room for the Desire to Receive for Oneself Alone to become manifest.

Rabbi Isaac Luria (the Ari), posits that the Lightforce pierces each and every stratum of existence, that the Emanator is all-embracing, that every molecule, atom, sub-atomic particle in the universe, every cell is imbued with the power of the *Or Ein Sof* or Light of the Endless. The Lightforce is within us and without us, a part of us, yet apart from us. Everything in the universe is but an aspect of one, living, breathing organism.

Some spiritual teachings place the Lord high on a lofty ethereal pedestal, far beyond the reach of humanity. In some religions it becomes necessary to die in order to "meet your maker." Other philosophies see the Lord as having created the universe before moving on to bigger and presumably better things.

The kabbalistic doctrine of "no coercion in spirituality" required that the Lightforce could not enter our space called the "vessels," unless we have removed our "Bread of Shame" by the *tikkun* process.[24] This, in turn, permitted man the free-will to either exercise Restriction or fall victim to the Desire to Receive for Oneself Alone. No longer could the vessels, meaning ourselves, partake endlessly of unearned benevolence. Thus was born the concept of empty space. Vulnerability, unfortunately, now became part of the human landscape of suffering and misery.

It is in this concept that at any given moment there is created a space and frame of individual vulnerability. The reason a person may incur a disease or set the wheels in motion for some mishap is that within that empty space—where the Lightforce is not physically expressed—the Darkforce and his fleet of negative energy-intelligences are the powerful invaders capable of ravaging our body or our environment.

The Darkforce and his armada are fueled by cosmic channels.[25] When an individual incarnation expresses vulnerability, that condition creates the precise empty space the death star fleet has been waiting for. Like a deadly scorpion, the death star seizes the opportunity and pulls the individual into darkness.

The lapse in the body's defenses that allows cells to reproduce into a life-threatening tumor originates at these intervals of vulnerability created by the empty space factor. The death star fleet inhibits the body's immune system from performing the function that it has performed successfully for many years. This same factor determines why one patient lives and another with the same diagnosis and treatment dies. It also determines why one person incurs a disease and another does not.

In addition to the incarnation factor, there are several other causes that contribute to the assault on the human body. The danger zones mentioned in my book, *Navigating the Universe* represent a continuous barrage of unrestrained, uncontrolled energy-intelligence upon the human body and mankind's environment. Unchecked, our lives are destined to failure, misfortune, and chaos.

Weaving through the *Zohar* and its intertwining tales and narrations, we become aware of our whole environment as being engaged in a gigantic cosmic dance. We know that the Earth's atmosphere is continually bombarded by cosmic rays and cascades of energy coming down from outer space, destroying and creating in a rhythmic choreography of cosmic energy.

The future of man hinges upon the way the sky treats us. Man's spirit is almost totally dominated by an Upper, Celestial World of seemingly infinite resources. Burning heat, a greenhouse effect, lightning, tornadoes, hurricanes, to mention just a few, represent an unpredictable barrage of devastation beyond anything man could dream up for himself. The idea of celestial superiority is a part of everyday life. The sun and moon exert a profound and strong influence on our lives. The moon causes the sea to rise and fall with the tides. The sun keeps us warm in the summer and fades in the increasing cold of the winter.

Stretching the idea of cosmic intrusion one step further, the *Talmud* states that invasion by an evil eye is the principal cause of many sudden deaths without apparent reason.[26] We have already explored the *Zoharic* point of view regarding the evil eye and the significance attributed to it by the Biblical Code.[27]

In contrast to the mechanistic western view, the kabbalistic world view is one of organic unity. For the kabbalist, all things and events perceived or acting upon each other are considered as energy-intelligences that always remain interrelated and bound up with each other. While they appear as different aspects or manifestations within our world, they are considered essentially as part of the one unified whole. Our tendency to divide the physical and metaphysical worlds into separate concepts is seen by the kabbalist as something connected to the realm of illusion.

In the kabbalistic view, the channels of our five senses—the eyes, ears, nose, mouth, and hands—are intrinsically dynamic in nature. The kabbalist has come to see these functions as acting outside of ourselves in addition to providing us with the opportunity to experience the world around us. Our five senses are energy-intelligences, like the brain, that act upon our environment as well as being influenced by what is out there.

It is interesting and perhaps not surprising Rabbi Isaac Luria (the Ari), constructed an intellectual map of sense reality in which four of the five major senses are referred to as basic and are reduced to their general outlines of energy-intelligences, as follows:

> *There are essentially four basic foundations [energy intelligences] in each and every existence. They are sight, hearing, smell and speech. This is the secret of the four letters of the Tetragrammaton. These channels are motivated by the four energy-intelligences of the Lightforce known by their code names, "Soul of Soul (sight), Soul (hearing), Spirit (smell), and Crude Spirit (speech)."*[28]

Abstraction is a crucial feature of kabbalistic teachings. Because we tend to become involved in the structures and phenomena of our physical reality, we cannot always take all its features into account. Our knowledge is customarily a system characterized by the linear, sequential structure, which is typical of the way we think. What seems to emerge from the Ari is a phenomenon that should be dealt with when considering human behavior: Body energy-intelligence.

The eyes, states the Ari, symbolized by the *Yod* of the Tetragrammaton, portray the most significant aspect and intensity of the Lightforce. Consequently, the energy-intelligence generated by the eye can easily be transmitted to another person or thing. If, for example, the energy to be transferred is of a negative nature, and the individual receiving this transmission has no security shield set up or has, at that moment, a breach in the immune system, he or she is then vulnerable to attack. At that very moment, whatever is airborne will infiltrate the immune system, lock into its system and prepare its strategy for the ultimate kill.

This scenario is very similar in behavior to the deadly heat-seeking sidewinder missile that locks onto an enemy fighter plane and then destroys it. Once the deadly energy-intelligence becomes an integral part of the whole person, all hell can break loose depending on what is hovering in the cosmos.

The power of the eye, when understood from a kabbalistic perspective, becomes as powerful a healing instrument as it can be a devastating channel for destruction. Laser technology is rapidly overtaking conventional surgical procedures. To heal a diseased or broken bone, one must reach into the area. Laser can penetrate and invade the body without incision or surgery. This power and penetration of sight is not exactly a new prescription. The Bible and *Zohar* are replete with narrations concerning the awesome power of observation and evil eye.[29]

The questions that must be raised at this time are: How and why does the transmission by an evil eye create such misfortune? Where does the aspect of evil eye originate?

> *"...and shall send him [the goat] away by the hand of a man that is ET."*[30] *The word "ET" (in readiness) contains a hint that for every kind of action there are men specifically fitted. There are some men specially fitted for the transmission of blessings, as for instance a person of a "good eye." It is written concerning the priest: "A good eye can transmit positivity and blessings"*[31] *for through the priestly eye, blessings and healing become manifested. There are others who are specially fitted for the transmission of negativity and curses. On whatever their eyes fall their curses are confirmed....*

> *Hence... a man should turn aside a hundred times in order to avoid a man with an evil eye.... The priest was able to recognize such a man because he had one eye slightly larger than the other, shaggy eyebrows, bluish eyes and a crooked glance.*

> *... In Gush Halba, there was a man whose hands brought death to whatever they touched, and none would come near him. In Syria there was a man whose look always brought misfortune, even though this man meant it for good. One day a man was walking in the street with a beaming face when this man looked at him and his eye burst.*[32]

What seems to emerge from the *Zohar* is the power of the eye and its ability to attack another person by projecting negative energy-intelligences. If someone you know thinks negatively of you, the eye channeling can affect your physical and mental well-being. The energy-intelligence of a negative eye can stretch far and wide so as to create mishap in one's life by accidents or otherwise.

Then there are others who may not intentionally wish you harm but because of their own particular lack will project this evil or wanting. For example, a man who is childless, upon seeing the children of his brother, may mentally focus upon those children the deficiency thought-energy-intelligence he experiences. Children are generally considered to be more vulnerable to negative thought-energy-intelligences, and most do not realize their vulnerability.

The reason for the weakness in children is the absence of their Desire to Share, which only becomes an integral part of the internal soul at age 13 for a male and 12 for a female.[33] Until such time, the Desire to Receive for Oneself Alone has dominion over children. Consequently, they draw to themselves any form of energy out there, since they are principally takers. At maturity, when an energy-intelligence of the Desire to Receive for the Sake of also Sharing becomes a participant in their internal psyche, their whole mental, internal structure becomes essentially balanced until such time when negative activity enters into their personalities.

However, this form of vulnerability is not limited to children alone. We must always keep in mind the cosmic danger zones stated in *Navigating the Universe*, when and if we are vulnerable due to some negative activity. Therefore we must be overly careful of our actions during periods of intense cosmic negative influence.

Another area of unseen influences relates to what I refer to as the environmental energy attack. The world view that lies at the foundation of our culture has to be carefully re-examined. It was Galileo Galilei, a 15th century Italian astronomer, physicist, engineer, philosopher, and mathematician, whose work influenced the scientific revolution of the Renaissance, who proposed that in the true science, the scientists must restrict themselves to studying only the essential properties of material entities. Only those factors that could be measured and quantified could and should be dealt with. Other features or characteristics, such as smell, taste, sound or color were to be considered as subjective mental projections that should be excluded from the province of science.

By directing our attention to quantifiable properties of matter alone, we have lost our touch with the 99 percent of reality that completely surrounds and envelops us. With the uncertainty principle injected within 20th century physics, we have come to realize that there is no absolute truth in science. The idea that all our theories and concepts are uncertain, limited, and at best only probable is still, unfortunately, not acceptable by the majority of scientists. The layman has all but relinquished any idea of the importance of self within the so-called hierarchy of science.

To the scientists, there is no purpose for internal energy in physical matter. Nature is merely a machine working according to mechanical laws without the participation of mankind. Their view of nature is that of a perfect mechanism with exact, rigid mathematical principles. The layperson's attitude to his environment became one of indifference. We are here for the sole purpose of manipulating and exploiting nature.

In the 17th century, Johannes Kepler, a German mathematician, astrologer, and astronomer, devised empirical laws concerning planetary motion by studying astronomical tables. In the 15th century, Galileo Galilei performed experiments to discover the laws of falling bodies. Then in the 16th century came Sir Isaac Newton, an English physicist and mathematician, with his sudden flash of inspiration when he saw an apple fall from a tree. In the 20th century, however, scientists faced a serious challenge to their ability to understand the universe. In their struggle to grasp the new reality of sub-atomic physics, they became painfully aware that their basic attitudes and concepts concerning nature, as well as their fundamental way of thinking, were completely inadequate to describe the essence of our natural environment. However, the average layperson has been left behind in this new awareness.

Western thought has programmed the individual to believe that the extrapolation of the laws and principles of our universe is the function of science and government, and that this is not a project for the individual. However, it is the individual who delivers the mail, not the post office; it is the individual who fails to consider the effect of pouring toxic waste into the atmosphere and waterways; it is the individual who carries on the acts that are environmentally beneficial or destructive, not the government agency.

The misunderstanding of the role of the individual in public affairs is demonstrated by the story recently told by a news commentator. When it was reported to a group of U.S. residents that the cost of guaranteeing the deposits of failed deregulated Savings and Loan banks would be about $1000 for each taxpayer, they commented: "Why should we pay for it? Why does the government not pay the cost?"

If our present attitude toward the condition of the environment and the socio-economic structure is one of helplessness and hopelessness, it is because during the past 300 years the individual has been excluded from the knowledge that is needed and has surrendered responsibility for his or her actions.

The argument may be presented: What difference does it really make whether or not we understand our environment? One merely has to look about and see what remarkable achievements science has made and how much better off we are. Unfortunately the reverse is true. With the lack of experience and education in dealing with the effects of our actions, a feeling of helplessness pervades and is combined with the hope that the "authorities" will take care of the mess.

According to scientific thought all that was necessary was domination and control over nature. This control was to be left to the scientific establishment. The average man need not be concerned, nor was he capable of forming a system that could provide him with an organizational structure of order in the universe. Life experiences are hard proof evidence of the failure to provide a society free of violence, chaos, and disorder.

The drug problem spans the entire socio-economic spectrum. Doctors, lawyers, athletes, corporate executives, workers, students, housewives—young and old, rich and poor, black and white—no one, it seems, is immune to this all-pervading menace. Staggering numbers of people from all walks of life are addicted to prescription drugs, uppers, downers, valium, diet and sleeping pills. Millions smoke marijuana or take cocaine on a daily basis. Alcoholism continues to be an international disgrace. An estimated 365,000 people die each year in the United States alone of diseases related to cigarette smoking.

As is the case with so many of the predicaments that western societies face today, no one, it seems, is even considering, much less addressing, the cause of the drug problem and other epidemics. Instead we look for symptomatic solutions. The stress factor is commonly used as the cause for most of the world's ills. This reason alone does not explain away why people in a similar position of stress or other symptomatic factors are not vulnerable or exposed to the same hazards as the "unlucky ones." In the final analysis, it appears that we must constantly rely on whether one is lucky or unlucky. This reasoning obviously leaves little, if any, room for free-will.

Primal cultures have always practiced methods of transcendence, songs, dances, and ceremonies by which they achieved a stress-free society. They were capable of creating security shields that would protect them. What was once a rich social, cultural, and spiritual fabric has been eaten away by empty material concepts, false technological promises, and finally digested by the great, blind myth called "progress." Today, that once intimate relationship we had with nature has been replaced by crass illusions provided by a cornucopia of so-called recreational drugs.

The kabbalist has long known that our thoughts shape what we perceive as reality every bit as much as our reality shapes our thoughts. We are what we think. More than simply a means of perceiving reality, our thought-action has the ability to create the reality we perceive. We are more than observers of reality, and we are even more than participators in our Earthly conception of what is "real."

Action, according to kabbalistic wisdom, not only determines the Earthly reality we choose to create, but also molds the way in which we choose to interact with it. A classic demonstration of this frightening, unlimited, yet controlled power is described in the *Zohar*[34].

> *Rabbi Chiya and Rabbi Yosi were strolling in the wilderness. They noticed a man coming towards them. Rabbi Chiya said, "Let us turn away lest he be a celestial idol worshipper or ignorant one, and it is forbidden to associate with him while in journey." Rabbi Yosi said, "Let us remain here and observe, possibly he is a great wise man."*
>
> *As the man approached them, he said, "The path you intended to take is a dangerous one and I fear to*

*go there by myself. I know another way. I also felt
that I must warn you, for it is written, 'Before the
blind, do not place a stumbling block.'*[35] *Let us
immediately turn away from this road, and no
discussions until we are far away from its influence."*

*After leaving that place, the man explained his
actions. "That road is dangerous for anyone who goes
by it. Once a wise priest walked together with an
illiterate, ignorant priest. The ignorant priest arose
against him at that place and killed him. From that
day on, anyone going by that road places himself
in jeopardy. That place has drawn robbers and
murderers who lie in wait for passersby, and fall
upon all that venture by that way, to rob and kill
them. The force is now manifested and requires the
blood of that priest every day."*

What seems to emerge from this startling *Zohar* is that the
mechanistic paradigm of nature has to be abandoned in favor
of giving consideration to the internal 99 percent level. The
Cartesian method has indeed brought spectacular success in
the 1 percent area of the material level, the level with which
we can all associate and easily accept as reality, to the exclusion
of the more important realm of the internal 99 percent level
of existence.

This is precisely the reason why we have had very little success
in the areas that count more in importance. An individual
afflicted with heart disease or cancer is more than willing to
forsake the material luxuries and successes that science

brought and replace them with a better understanding of health and illness.

This idea is clearly stated in the *Zohar* when it declares:

> *When there is separation of thought between the internal (Zeir Anpin) level of existence and its external (Malchut), trouble and great pain reign in the world. When there is no separation, perfection, peace and harmony are dominant.*[36]

Therefore, our responses to the environment are determined not so much by the outer, detected, visualized effect of external stimuli on our mental and biological system but rather by unseen but very real influences. In the traditional Cartesian view, it is assumed that we all have basically the same biological apparatus and that we all have access to the same frame of sensory perception. This was and still is a basic flaw in science. There are those individuals who can instantly sense and distinguish the vibrations of an environment, whereas others draw a blank.

If only we listened to the words we use each day: "What do you feel?" and "Tell me what you see?" Both expressions usually do not refer to the obvious, external aspect of the subject under consideration. Returning to the narration of the *Zohar*, the point being made is that inanimate objects are not always lifeless, inert or unconscious. The Cartesian view of matter as something that we have control over must be stretched beyond the physical limitations of science. Our thoughts, our actions are sucked in by all material entities around us, both on a terrestrial and celestial level. These so-called vibrations that some of us intuitively sense are the energy-intelligences of human beings that have become an integral part of the subject, whether it be a road, home or place of work.

Thus we come to the realization that there are essentially two realities. The material 1 percent reality lends itself to a reductionist description, adhering to the Cartesian method and viewpoint that everything is like a machine, constructed from separate parts. The 99 percent internal reality, unfortunately, has not been deemed worthy of scientific investigation. This reality is what most contemporary scientists find hard to admit. In fact investigation of this most important 99 percent aspect of the universal whole is not encouraged and questions concerning its existence are hardly ever raised.

Because the reductionist approach is inappropriate for solving the more serious problems of mankind, no dramatic solutions have come forth. How, then, is this situation going to change? We must go beyond the narrow mechanistic framework of contemporary science and develop a broader approach to problem solving. Transcending this limited view of our universe will require a major cultural revolution.

However, in this Age of Aquarius, Rabbi Shimon Bar Yochai has acknowledged and forecast this information revolution. The new framework formulated by the *Zohar* will not only have a strong impact on our lifestyles but will also have the potential of being spiritually and physically unifying. To develop a kabbalistic approach for the improvement of our personal and environmental well-being, we do not need to break completely new ground. We can integrate *Zoharic* knowledge with the basic laws and principles established within the scientific community.

Modern scientific thought must lead to a view of reality that comes close to the views of the kabbalist, in which knowledge of the human mind and body, together with accurate celestial data concerning its unseen influences, become an integral part of our lifestyles. In addition, we must achieve an awareness that our entire universe is in a natural state of dynamic balance.

The human organism is the decisive factor. All celestial bodies behave in particular ways because they are endowed with intrinsic natures that make their behavior inevitable for them. Kabbalistic astrology probes the very nature of reality. Experimentation was once an attempt to study a system through analysis based on controlled stimulation. This exploration was then followed by observation of the resulting response. However, ever since Werner Eisenberg's uncertainty principle,[37] controlled experiments on a cosmic level could never be accurately determined. Thus, results of calculations by theoretical astrophysicists seem almost pointless. One merely has to reflect on human activity to suspect that some mad metaphysical scientist is on the loose. In fact, according to Kabbalah, the chaos and tumult of physical existence was actually placed on Earth for the specific purpose of allowing man free-will sufficient to alleviate Bread of Shame.[38]

The Kabbalah teaches that before there was the wheel, there was the idea of the wheel. Not only do thoughts and ideas enable us to create the physical world, these same thoughts have influence over that which occurs in the cosmos.[39] We understand that the moon affects the tides. We acknowledge that supernova, black holes, and other phenomena in outer space inevitably affect weather and other conditions here on Earth. But can we comprehend the ancient kabbalistic belief that the behavior of Earth-bound people can override extraterrestrial influences and, contrary to scientific belief, even have sway over intergalactic events?

The destruction by the Romans of the Second Temple in Jerusalem in the year 70 CE brought the near disappearance of answers to such questions. Through the long centuries since that time, the light of Kabbalah flickered, but this vital ancient wisdom could never be extinguished. It is written in the *Zohar*[40] that Kabbalah would have to await the coming of the Age of Aquarius to make its reappearance as an instrument to be wielded by the hands of man, as the electronic means to draw the Lightforce upon a human race, wandering and confused in cosmic darkness.

That time has come. The *Zohar* is a book of power, the power to form the letters of the *Alef Bet* and to make them do one's bidding. But the *Alef Bet* is of no practical use if we do not understand how to patch ourselves into this supreme, all-encompassing network. The study of Kabbalah allows us access to this life-sustaining system.

Chapter Two

THE MIND-BODY CONNECTION

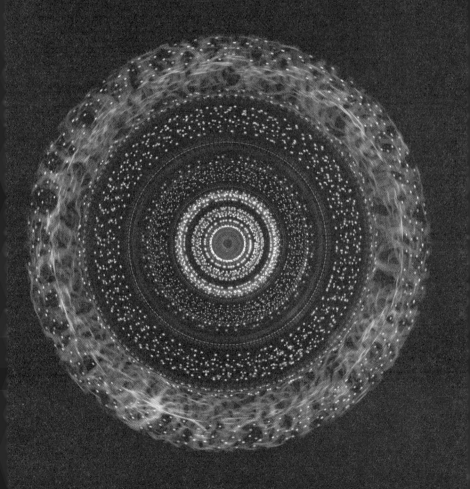

THE MIND-BODY CONNECTION

Then the eyes of the blind shall be opened,
and the ears of the deaf shall be unstopped.
Then shall the lame man leap as a hart.
And the tongue of the dumb shall sing.
Isaiah 35:5-6

THE PROMISE FROM THE BOOK OF ISAIAH WILL BE
FULFILLED with the ongoing Age of Aquarius. This
assurance was clearly made by Rabbi Isaac Luria in his *Gate of
the Holy Spirit* on the subject of healing.[41]

> *To remove the illness, one must take upon oneself*
> *the bitter conditions of healing... for the purpose of*
> *grasping and understanding the metaphysical*
> *teachings, which are the secret doctrines of the*
> *world. This is the wisdom that has been concealed*
> *from the early days of Rabbi Shimon Bar Yochai*
> *until now (1572) and as the Rashbi (Rabbi Shimon*
> *Bar Yochai) stated, permission shall not be granted*
> *concerning its revelation until the final generation*
> *that will usher in the Age of Aquarius, whose time*
> *is now, through the medium of the Saintly Teacher*
> *Rabbi Isaac Luria, with the assistance of the*
> *prophetic spirit within him.*

The human nervous system is the most complex physical structure in the fabric of man. Infinite interconnections and electrical impulses permit us to think, act, and create, and more importantly, to understand who we really are. Vast research has been directed towards the connection between mental activity and our physical body.

Research indicates that the mind actively participates in curing sickness. Psychic imbalance is considered to be the root of all illness. Consequently, there emerged the popular contention that stress is the fundamental and primary culprit to be dealt with when improvement in mental or physical well-being is to be achieved. We have already addressed ourselves to this phenomenon with the conclusion that stress and other contributing factors are not the primary cause of the universal ills that befall mankind.

René Descartes' 17th century philosophy of strict division between mind and body led the medical establishment to concentrate on the body machine. Environmental factors were neglected. The trend now among a strong minority of scientists is towards a mind-body connection. Their findings, for example, are of critical importance to persons with so-called degenerative diseases. They suggest that the effects of emotional stress can suppress and even weaken our immune defenses.

We also intuitively recognize the existence of a connection between mind and body. Generally speaking, it is not as easy to perceive the link between mind and space. The ancient axiom, "as above so below" alludes to this cosmic connection.

From a kabbalistic perspective, all of Creation is part of the All-Embracing Unified Whole. Thus, as in quantum, it only stands to reason that anything occurring anywhere instantaneously influences everything else.

Kabbalists have always engaged in what has popularly come to be called "the power of mind over matter." They take this concept one step further than the quantum physicist. Kabbalists suggest that more than being a mere participator in the metaphysical (quantum) scheme, man, utilizing the power of thought, can act as a determinator of both physical and metaphysical activity.

The awesome technology of progressive medicine creates an image of such great potency, with its massive structures and knowledge, that individuals find it hard to believe that they can make much difference or accomplish significant changes in their own well-being. The achievements of modern medical science is by no means to be down-played. But in the future, holistic techniques that include the cosmos and mind will play a greater role towards achieving human cosmic well-being.

The kabbalistic conception of mind over matter does not necessarily correspond entirely to the popular connotation of that subject. Telekinesis, for example, is the physical movement of objects through the power of thought alone, such as bending keys, stopping and starting broken watches. While it is certainly within the realm of practical possibility, it is not, to the kabbalist's way of thinking, of worthy pursuit. The reason is that to engage in such activities is to play, so to speak, into the hands of the Cartesian paradigm.

What, after all, is the purpose of bending a key or guessing symbols on a card, if not for self-aggrandizement, entertainment or to prove to some so-called objective observer the power of mind over matter? A far more productive use of thought-energy, claims the kabbalist, is to power the mechanism by which we become engaged with the Infinite Reality. For by doing so we then come into control of our destiny and the whole of the cosmos.

When the kabbalist deals with the concept of mind over matter, we must understand that he or she is speaking of undergoing an alteration of consciousness. What is necessary is a transformation of the mind from the rational logical mode of five or seven percent of our potential to the cosmic mode that allows for the conscious transcendence of physical limitations and restraints.

Thought can traverse great distances, can affect people and objects, and is indeed a tangible factor in the world around us. It is no fault of Kabbalah that traditional science cannot, as yet, recognize or understand this. It is possible to remove oneself from the impelling negative influence of degenerative diseases of the body. The stars impel but they do not compel. And since the Kabbalah establishes the cause for all illness and mishaps as originating with negative energy-intelligences of the cosmos, the task of the kabbalist is to rise above these negative influences.

Essentially this requires the individual to detach the self from the confines of physicality and connect to the positive energy-intelligence of celestial bodies. Humanity's finite aspect, which may be described as the flesh and bones, is subject to Cartesian rules and regulations. However, the other, the infinite characteristic, operates beyond limited physical jurisdiction. The finite is subject to pain, discomfort, and death. The infinite is in the eternal realm.

When we connect with our infinite aspect, namely our soul, we must also pay constant homage to the original act of Creation, namely Restriction.[42] Through this channel it becomes possible for us to transcend space, time, and matter. We also receive the added potential for astral travel and the instant alleviation of physical and mental pain and suffering.

Physically, we are creatures of Earth; spiritually we are perpetually connected to the Endless.[43] The finite part of us is subject to change, turmoil, pain, and suffering. The other higher aspect remains beyond the jurisdiction of physicality.

Through the kabbalistic attitude of positive resistance, a connection can be made by which the Infinite Self is illuminated. By connecting with the Infinite aspect, a transformation of consciousness takes place that allows us to rise temporarily above the time-space continuum, beyond pain and physical discomfort, above the machinations of the physical world.

According to the *Zohar*,[44] the day is nearing when the inner secrets of nature that have remained so long in hiding will at last be revealed. Such knowledge will enable us to reach the very essence of that which is of us and around us. This knowledge will permit us access to the non-space domain. It will provide us with a framework for the comprehension of not only our familiar, observable universe, but also of that which lies beyond the range of observation in the realm of the metaphysical.

The laws of cause and effect are valid in the spiritual as well as in the material world. We have no trouble identifying that which is material. It has body, substance. We can touch, see, smell and even hear it. But when we get down to the "substance" of what we perceive as solid, physical reality, we find the basic building block of nature to be the electron.

And what is an electron? Is it a microscopic bit of solid matter? Not at all. It does not even occupy a specific place in space/ non-space. Thus we find that the fundamental property from which the material is constructed is an illusion. All that remains is that which we, at this moment, are sharing: thought energy-intelligence, the unique, particular life form.

In 1930, scientist Sir James Jeans summed it up when he wrote:

> *"The stream of human knowledge is heading towards a non-mechanical reality. The universe begins to look more like a great thought than a great machine. Mind no longer appears to be an accidental intruder into the realm of matter. We are beginning to suspect that we ought rather to hail it as the creator and governor of this realm. It may even be that what we think is the real and the physical universe is just an interference pattern (an impertinent blip) in the world of thought."*[45]

This startling revelation might well have come from a madman or inmate of a mental institution. Does Sir Jeans have us believe that, in reality, there is no other reality than thought? When he addresses our physical reality, he considers the physicality itself to be the universe's handicap, interfering with our daily life. Our entire environment of physical pain and disorder is dismissed as a distortion, and in essence, it is not even to be considered.

Surprisingly enough, the kabbalist joins in and completely agrees with Sir James Jeans, with one further consideration: The idea of reality is not something directed only towards outer space or the universe at large. The kabbalist directs this awesome revelation to man himself. The physical body interferes with his own thought processes, often completely obscuring them.

This interference is the power of the Desire to Receive for Oneself Alone, which can be discerned as the material or body energy-intelligence. The energy-intelligence of the spirit is known as the Desire to Receive for the Sake of Sharing. It is the material force that creates havoc within our universe. Kabbalistic teachings stress that this is the fundamental and essential cause of life's problems and mishaps.

The universe—and man in it—is but an enormous composite of thought. The universe exists in our minds and in our bodies, in everything we experience, taste, touch, see, and do. Every force observed, whether particle, anti-particle, neutrino or quark, is directed by and acts according to the dictates of a particular intelligence of thought. Thought has been quantized by science as a discrete pocket of measurable energy, but in fact, it is a part of the all-pervasive whole.

The forces at work in nature, known to Kabbalah, are independent of time, space, and motion and can best be considered as ongoing permeations or states of being. Let us be content at this stage of our investigation to think of all energy as being created in thought and as states of mind.

What seems to emerge from all this is that all that exists is thought, from a dinner table or home to an electromagnetic energy field operating in space or non-space, in time or non-time, illusion or reality. Like that of the Lightforce, in whose image we were created, the mind of an individual is not only where information is stored, it is where energy-intelligence and knowledge are created.

Wireless broadcasting networks are not at all original to the 20th century. Consciousness acts upon energy-intelligent thought and transforms it into energy-intelligent matter. Through each mind's own unique perception, a new concept is programmed into the universal grid that is instantly flashed to the minds of all our fellow inhabitants. The instrument by which consciousness performs this miracle is electronic. It is the Hebrew *Alef Bet.*

Therefore, at the table in the restaurant where we are now sitting, the former occupants have instilled energy-intelligence and thought consciousness which may be positive or negative. When we rent or purchase a new home, we must recognize that the energy-intelligence and thought consciousness of its previous residents are permeating this space. Were they positive or negative type people? To most of the readers of this book, this may sound downright absurd, at its best, and idiotic, at its worst.

However, I have made an attempt to crystalize this radical approach to the reality realm so that we can all become aware of the unseen influences that affect our very lives and well-being. Our conscious attention is generally engaged by the physical objects and thoughts pertaining to the activity of the moment. In addition, however, unseen influences are having an effect on our thoughts and behavior without the benefit of our conscious evaluation. Kabbalistic teachings offer an opportunity for the individual to become increasingly aware of the unseen influences.

These unseen influences are real and are very much a part of our human landscape. They penetrate the barriers that have been placed within the universe for our protection. Whether we call these security shields, immune systems, ozone layers or other devices established by the Lightforce to balance our universe, these shields can be penetrated at a precise moment in time when cosmic danger zones[46] are present or when the world security system has been pierced by the Dark Lord and his death star fleet. At that moment we are faced with the possibility of becoming vulnerable to their onslaught.

Neither a religion nor a philosophy, kabbalistic teachings are a way of life, a natural technique for the restoration of our security shields and accomplishing an ultimate balance within the cosmos. These teachings are by no means effortless. The requirements of Restriction and the Desire to Receive for the Sake of Sharing are very demanding.[47] No pain, no gain.[48] The more pressure a runner undergoes, the better athlete he becomes. Lifting heavy weights strengthens the body. On a metaphysical level, we must also work to control pressure and stress so that we may be in harmony with ourselves. The lack of effort creates a space for the Desire to Receive for Oneself Alone to reign supreme.

Stress alone is not and cannot be a factor in the breakdown of our security shield. Neither is the activity of the mind by and within itself a problem. What seems to be more important is our vulnerability. We must remember that thought energy-intelligence is our connection to the Lightforce.

It is important that we achieve an awareness of the unseen influences around us. We must understand how the mind, thought energy-intelligences, can assist in promoting the welfare of the universe in general and mankind in particular. The *Zohar*[49] gives us a striking demonstration of how unseen influences affect our way of life with its interpretation of an unusual abstruse section of the Biblical Code.

> *Similarly, it is written, "When you come into the land of Canaan which I give to you for a possession, and I put the plague of leprosy in the house of the land of your possession."*[50]

> *Now when the women brought articles for the Tabernacle, they used to specify what part each was for: "This is for the Holy place, this is for the*

curtain."[51] *Similarly, whenever anyone makes something for idolatrous worship or injects negative energy-intelligence short-circuitry, as soon as a person connects his or her negativity with that "thing," a spirit of negativity rests upon it.*

Now the Canaanites were idolaters and whenever they began to contract a building for their fulfillment of evil they would not only think evil but also utter negative energy-intelligence. Thus a spirit of negativity would rest on the building.

Therefore, when a person begins to set up a building, he should declare that the building is also for the service of the Lord. Then the Lightforce is drawn into the building and bids peace be with him. This is indicated in the verse, "And you shalt know that peace dwells in your house."[52] *Otherwise, he creates a space for vulnerability for negative energy-intelligence. Then indeed, a spirit of short-circuitry will rest upon the home and that person will suffer the consequences, and whoever dwells in it may come to harm.*

If it is asked, how is one to know such a house? It is one in which the man who built it has come to harm, he or his family, whether through sickness or loss of money, he and two others after him. Better a man should fly to the mountains or live in a mud hut than dwell there.

> *Therefore the Lord revealed to the nation of Israel that which could not be known [on a corporeal, material level, these negative energy-intelligences could not be observed]. It is written: "They know not the manner of the Force of the Land"*[53] ... *"And he shall break down the house, the stones thereof, and the timber thereof."*[54]

> *We may ask, since the uncleanliness has gone, why should he break down the house? The reason is that as long as the house stands, it belongs to the "other side" and it may return.*

The lengthy *Zohar* portion just quoted reveals for us some remarkable and astonishing insights into the world of reality and metaphysics. The *Zohar* places a great deal of emphasis on the awesome power of thought, to the extent that inanimate objects are also subject to thought energy-intelligence. Furthermore, the idea that these inanimate entities can affect our health and environment is clearly stated in the *Zohar* based on material furnished by the cosmic code of the Bible. Misfortune and illness are not considered separate and apart from our thoughts. Our mind and body, our thought and environment are inseparable.

Therefore, not just the disease itself must be treated but the entire self and the thoughts of others. Our state of mind can make ourselves and others well and can speed recovery from illness. We have the power to heal and the power to remain well.

Perhaps one of the most dramatic cases that illustrates the power of the mind over body has been reported by Dr. Bruno Klopfer, a psychologist and researcher involved in the study of the drug Krebiozen. In 1950, Krebiozen had received sensational national publicity as a "cure" for cancer. The drug was being tested by the American Medical Association (AMA) and the United States Food and Drug Administration (FDA).

A patient had developed advanced lymphatic cancer, a generalized, far advanced malignancy involving the lymph nodes. The patient was included in an experimental study of the since discredited drug Krebiozen. The patient had huge tumor masses throughout his body and was in such a desperate physical condition that fluid had to be sapped from his chest every two days. When the patient learned about the drug Krebiozen, he begged to be included in the study.

After one dose, his tumors almost disappeared and the patient's recovery was indeed astonishing. The patient regained enough strength to resume a normal life. When the first published reports of the AMA and FDA claimed that the drug was ineffective, the patient took a dramatic turn for the worse. The tumors grew and he once again became bedridden. In a desperate attempt to save him, his physician told him that the reports were false and that double-strength doses of Krebiozen would produce better results. Actually, the injections consisted of sterile water.

The patient again experienced rapid remission and once again the tumor masses melted. Soon he even went back to his hobby of flying. When the Food and Drug

Administration announced its final findings, which appeared in the media as follows: "Tests conclusively show Krebiozen is a worthless drug in the treatment of cancer," the man died a few days later.

How can the placebo effect be explained? Some dismiss the phenomenon by attributing the illness to a psychosomatic process. Some say it is a figment of one's imagination or another way of saying the same thing, "It's all in the head."

However, taken at its root meaning, the word "psychosomatic" means that a medical problem originates and is continuously worsened by a person's internal mind or psychological process. We cannot disregard any sickness as not real simply because its origin is not in the physical realm. This idea of psychosomatic connections, while still generally accepted, is not new and does indeed exist. Scientists are only now beginning to map and trace the stress routes from the brain to other parts of the body.

Consequently, our mind acts as a healer as well as destroyer. Taking this idea one step further, the kabbalistic world view suggests that the mind can extend its influence over the entire cosmos. The *Zohar*, mentioned previously, clearly states that man has control over the kingdom of the inanimate. Thus cosmic influences, which lie at the heart of all misfortune and illness, should be subject to human control, and thus these influences can be made to behave in accordance with mankind's directives. Kabbalistic teachings demonstrate how people can exercise substantial influence over bodily states that were formerly considered not to be subject to conscious control.

The basic framework of this book is directed towards strengthening the connections between body, mind, thought, and the cosmos. Verification of these phenomena has been left to research scientists. The task of scientifically validating these observations has been undertaken by researchers the world over.

Another aspect of kabbalistic thought is that it seeks to address the prevention of vulnerability. While medical research has not as yet come up with adequate explanations of what causes most degenerative diseases, part of any investigation must ultimately address itself to another cause, namely vulnerability. We have all at some time, experienced a suppression of the body's natural defenses against disease, namely the immune system.

To understand the illness, we have to consider not only what causes the disease, but also why most people are able to avoid disease in the first place. We are all open to all sorts of diseases. This, however, does not mean that we will become sick. The body's defense system is so powerful and effective that most people exposed to all sorts of infectious diseases maintain their health. This is the serious dilemma facing medical research. In one case, the body does battle with foreign substances and subsequently destroys them. In another, with the same self-healing system, the body's defense mechanism fails to fight and devastate the hidden, internal enemy.

Vulnerability, is the kabbalist's explanation. The cosmos at definite time frames, attacks and suppresses our natural defense mechanisms. The body's immune system, which keeps close tabs on any abnormal cells and then homes in for the kill, can be inhibited by negative cosmic influences. The important point here is that something is happening, an unseen influence, that creates susceptibility. Kabbalistic teachings show us how to prevent moments of vulnerability from occurring within the body's defense system.

Let us turn for a moment to environmental hazards. The kabbalistic perspective of our universe runs a course similar to that of the new age of quantum mechanics. If we refrain from acting negatively towards nature, we are in harmony with our environment, our physical surroundings and the cosmos, as well as with our fellow human beings. The survival of our whole civilization depends on whether mankind comes to recognize that human activity strongly influences our whole environment.

However, what if others do not feel or see the necessity for improving our natural and cosmic atmosphere? How can I prevent the influence of their negative activity from affecting me? To achieve a dynamic balance with our environment, the Kabbalah teachings create the necessary precautionary measures by which we are neither influenced nor become vulnerable to its negative stimuli.

Our responses to the environment include our participation. We play an important and active role in restoring a dynamic state of balance. We can assure ourselves that negative energy-intelligences do not invade our space, whether that space be our body or the highway we are driving along. The invasion takes on many shapes, be it in the form of an enemy of our physical or mental well-being, the encounter with a drunk driver or our own dinner table.

To arrive at such a complete picture, the kabbalists not only developed highly refined diagnostic analyses of our cosmos and its environment but also a unique art of Kabbalistic Meditation that permits the connection of mental activity with the physical body and the universe. The brain's *13 billion* interconnected cells make it virtually impossible to trace the exact circuits through which consciousness operates. Although physically the brain amounts to only a few pounds of matter, its knowledge and information capacity, along with its unique switching capability, far exceed the largest computer.

Despite the many scientific advances, a vast and perhaps unbridgeable gulf still exists in our understanding of the physical process of the nervous system and thought consciousness. While we can investigate certain connections and correlations between physical phenomena and the mental processes, the nature of the link between mind and matter remains a mystery unpenetrated by scientific inquiry.

There is no question that the complex mechanisms functioning on the atomic, cellular, brain levels stagger the imagination. It is highly pretentious to assume that the mystery of thought consciousness will ever be unraveled by conventional analytical methods. Perhaps, one day, scientists will eventually have models of how impulses create thought, such as a mother's thought of "How wonderful and lovable is my baby."

Many scientific investigators of the brain and nervous system have finally come to realize that there is a quality of the mind-brain that transcends the biological. Though contemporary science has discarded the dualism of body and mind, the investigation of the brain has left it in awe of the mind.

How does the Kabbalah view the mystery of the mind? The brain and body are physical substances and are physically connected. Here the medical and psychological sciences find it easy to account for the effect of the mind-brain on the behavior and functions of the body. Most researchers conclude that the mind and all of its functions, such as consciousness and thought, are in actuality nothing more than integrated combinations of the brain's physical activities. They give rise to memories, perceptions, and the ability of nerve cells and neurons to change with experience and mechanically execute modes of behavior impressed upon the brain.

These conclusions are, however, mere speculations for there is no evidence that the origin of the mind is in the functioning of the brain cells and nerves. What forever defies explanation is how information sensed by our nerve receptors—whatever that means—converge in brain substance to become the subject and substance of thought. Putting this aside for a moment, science, in general, treats the mind as untouchable, even beyond the reach of scientific investigation. The mind, they claim, is simply the product of mechanical activity of the brain. Science talks a lot about the brain and little, if anything, about the mind. They cannot explain *how* the mind functions, *how* to expand the mind, and *how* we can use it more efficiently.

Mental or mind functions are not determined by the tidy, precise mechanics of the brain's myriad nerve connections. Science, consequently, can never conclude that it can account for every activity and phenomenon of mind consciousness. Herein lies the basis for the failure of medical psychiatry to resolve our mental health dilemma. Existing scientific information indicates that the mind is more than an entity that can be accounted for by the functions of the physical brain.

Despite years spent in the research of memory, the brain's ability to store information and recall it on demand still remains a mysterious phenomenon. The memory stores of the brain are filled with information from every experience in life. The operations of the brain in keeping the data in orderly sequence, arranged for relevancy, is startling in its complexity. So is the remarkable ability of the mind to have an awareness of events and things that have been experienced and then retrieve pertinent information about these things from its endless memory bank.

Like the most sophisticated computer, the mind has the ability to trigger the memory banks to produce a concept. Our conscious mind identifies some quality related to the concept and then commands and directs a search for the correct word or phrase. The concepts appear whole and coherent. Then follows other abstract phenomena of the brain-mind such as intuition, love, and loyalties. Then there are the unusual states of mind, such as dreams, illusions, internal sensations of peace, joy, and happiness. Who pushes the button and why at particular moments in our life? And how and why do individuals develop particular, unique, and different thinking styles?

Consequently, the kabbalistic perspective on mind-brain relationships appears as a refreshing fountainhead which challenges several notions about the origin and essence of the mind. The ideas initially may seem strange and foreign. Yet, if I am to be successful in making a serious and much needed contribution to this very important subject, these ideas must be presented. I will attempt to present them as clearly and concisely as possible.

A good starting point for our exploration into the world of the mind-brain is understandably the *Zoharic* view of this complex matter. As we shall observe from this perspective, it is suggested that our minds contain the equivalent of a "hidden" universe of activities. According to kabbalistic teachings, this realm of mind perceives vast ranges of stimuli from many sources. The mind causes whole cascades of involuntary physiological changes. The mind performs complex pattern recognition tasks and makes decisions that control just how much we know about what is going on around us. The mind will also determine what ambitions we wish to pursue and which not.

The mind even directs the events leading up to and including how wealthy or poor we shall become. For in the final analysis, failure came about because we did not recognize or take notice of some item that would have assured the success of the project. To some, whatever they touch "turns to gold," and for others "things never seem to go right." The pervasive illusion is that we dictate the scope and direction of mind conscious awareness. The reality is that the mind is *actually arranged by unseen forces* that operate to present to us an already structured situation, which we comprehend in its final, finished version.

From a kabbalistic point of view of reality, the story of our universe is really one of returning souls. Indeed, no mystery in the long history of our universe is so startling as the universal and repeated behavior of its inhabitants. This subject is so little understood that we need not be astonished at our continued insistence on destroying each other. If we must be astonished, let it be at our inability to unlock the secrets of human behavioral patterns.

Fundamental evolutionary precepts have hardly changed at all throughout history. We have witnessed civilizations weave through the fiber of recorded history, attempting to impose their kind of order. Yet the inevitable process of change that has become a by-word in high technology makes us wonder how basic life forms still hang on unchanged for so long a time. In a quickening evolutionary society, people, as well as all other life forms, maintain their desires for the very same things that prior generations did.

Startling breakthroughs that ultimately foster bigger leaps have had very little effect on human thinking. Conservative stability still remains the rule for most species within our universe. Is our frame of mind really different from that of the people of the Middle Ages? In spite of dramatic environmental changes, along with the progress syndrome, have human psychological needs really changed down through the centuries? Do we really better ourselves by growing with progress that continues to become more complex with the passing of time?

Consequently, when we come across information that can bridge the gap between the growth of progress and the lack of change in personal fulfillment, it is really exciting, and even more, refreshing. We therefore turn again to the *Zohar* in our attempt to throw some light upon the questions that have been raised:

> *Rabbi Shimon here introduced the subject of transmigration of souls, saying: "Onkelos translates the above words as follows: 'And these are the judgments which thou shalt order before them.'*[55] *In other words, these are the orders of the meta-psychosis, the judgment of souls, by which each of them receives their appropriate consequences [computerized cassettes]. Companions, the time has arrived to reveal diverse, hidden, and secret mysteries in regard to the transmigration of souls."*[56]

Thus, when considering man's behavioral patterns, we are in essence seeing aspects of ourselves in former lifetimes. Life for most of us is almost a re-run of our activities as experienced back in time, as well as tasks we attempted before but somehow failed. For precisely this reason, man remains in a fixed psychological state of mind. Therefore, man still maintains his primal characteristics and clings to well-defined modes of existence. Man, in the 20th century, is merely engaged in a mini motion picture played in re-run over and over again.

What emerges from the kabbalistic view, and what I am suggesting, is that although human behavior is genetically controlled to a significant degree, the *tikkun* process directs and dictates our everyday thought patterns, feelings, and activities.

Now I know that this position challenges the conventional view of most social scientists who claim that cultural and environmental upbringing, and not incarnation-related imperatives, shape human nature. The far reaching effects of our *internal human spirit* extend to our characteristics and determine our external actions, which are fully determined and executed by the cosmic forces prevailing at that time. Actions of man are indeed controlled by the cosmos but only to the extent that they were manifested in a prior lifetime. In other words, if an individual committed crimes against humanity in a previous incarnation, his incarnated soul returns and is faced with the same type of situation with which he was challenged in his past lifetime. He is now given an opportunity: He can exercise free-will and thwart the cosmic scenario, which determined and manifested the present life-existential cassette of a prior lifetime or he can succumb to its influence.

These negative frames of reference, *established by a former lifetime*, are manifested by the conglomeration of cosmic activities and their position in the cosmos at the time of a person's birth. In essence, the cosmos merely presents the opportunity and framework for our manifested incarnated cassette of lifetime activity. The cosmic strings of activity are not the cause of the life cassette's pre-determined structure. This has already been determined by our former lifetime. The set of circumstances occurring in our present lifetime is a result of the accumulation of cosmic influences coming together at this time and affecting our *tikkun* requirements, thus producing a unique Life-force cassette.

Is this arrangement really fair? The answer lies in the initial purpose for Creation, namely, the removal of Bread of Shame.[57] However, we might again raise the question, "What are the chances of succeeding this time around when we have already failed in countless prior lifetimes?" These negative frames of reference created by ourselves provide us with an opportunity to exercise free-will and achieve the removal of Bread of Shame. Obviously if these negative cosmic forces did not exist, then man would simply conform to a programmed kind of intelligence that dictates the sharing philosophy, leaving no room for passions, hatreds or other objectionable characteristics that distinguish us from robots.

Therefore, while on the one hand cosmic negativity arouses evil behavior, this influence, as powerful as it is, can and should be regulated and controlled. This is the obligation and purpose of free-thinking and free-choosing individuals. But as history has demonstrated, man has failed to accomplish mastery of his destiny. Now, because the teachings of the Kabbalah are being made available to all, man can overcome this failure.

A further question of interest is "Why now?" Human activity at present is unstable. With scientific advancements, we have become aware of internal, metaphysical activity that seems to create even more uncertainty. Yet, it may be that because of these scientific advances, we are becoming so much more enlightened that we *now* demand to know who or what is the cause of the apparent instability and uncertainty in our lives.

The influence of the Age of Aquarius will find our civilization preoccupied with information and enlightenment.[58] Although the Kabbalah has been a jealously guarded secret, the time has come for it to reach the masses. We now live in a time of enormous upheaval. A time in which the mores, the traditions, and the answers of the past are all in question. At this point in time, the power of technology and the success of science have created a situation in which it is very difficult to bring together the physical and the mental.

The human being may be seen as an information processing organism whose various traits are closely tied to the complexity of the data presented from an environment of former lifetimes. The mind-brain system is a complex system of interconnected, interdependent structures and functions projecting an already produced film. The mind-brain can process any information all at once, faster and more reliably than any present or future computer. The psychological and perception process involves *infinite* levels and stages of processing, a task never too difficult for our mental computer. Some of this involves infinite parallel and sequential processing, which the mind-brain never finds too straining or overwhelming.

The initial stages of processing always occur outside our awareness, it is a program that has already become finalized long before any consciousness processing ever occurs. It is for these reasons that kabbalists have always recognized the domination of robotic-consciousness over our ego-centric consciousness. Our conscious awareness, which plays *no role* in the initial and final stages of any process, has, for too long, enslaved us by convincing us that we are indeed in control of our fate, destiny, and decisions. The kabbalistic view, namely that incarnation and the *tikkun* process determine our behavior, is only recently becoming generally accepted.

The unconscious processes are far more pervasive than the conscious processing. The opinion that our mind-consciousness functions at only five percent of our potential has long been accepted in mind research. Consciousness is considered, if it is considered at all, as a later and sometimes *optional* stage of cognitive information processing.

Consequently, the crystal-clear message appears to be that fate, destiny, decisions, and human behavior cannot be understood without taking the unconscious psychological processes into account. No psychological model that seeks to interpret or explain how and why human beings behave, learn, or experience different sensations can possibly ignore the existence of unconscious psychological processes.

It, therefore, comes as no surprise that mental illness and its institutions continue to increase at an ever expanded pace, with no relief in sight. Rabbi Isaac Luria (the Ari), emphasized this kabbalistic doctrine, when he stated the following:

> *No individual can ever achieve a completed phase of teshuva, [a Back to the Future concept]* [59] *where the individual comes into total control of his fate and*

> *destiny, unless he becomes knowledgeable of the*
> *unconscious root psychological processes of the soul*
> *along with the knowledge of former lifetimes.*[60]

What seems to emerge from the Ari's writings is that soul-consciousness not only explains seeming paradoxes in our society but it also provides us with a new approach to the problems of universal mental illness. Its message is that no political, economic or national environment is responsible for our behavior or for the ills and misfortunes of our society. This important aspect of consciousness enables us to perceive how the whole pattern, the underlying paradigm of western belief in the understanding of the brain-mind leads inexorably to the kinds of problems we now face. It creates a belief that these dilemmas have their satisfactory resolution only through change in the dominant paradigm.

The *Zohar* is very emphatic on the importance of understanding one's internal soul-consciousness when it states:

> *Wisdom is the knowledge connected with an*
> *understanding of the soul. What does it embrace,*
> *where did it come from, why has it been directed to*
> *enter this body, which is an illusion? For today, the*
> *body is here and tomorrow in a grave.*[61]

The *Zohar* considers the answers to those questions necessary to back ourselves right up to the edges of reality and come face-to-face with the internal consciousness and Divine view of ourselves. Human consciousness has the capability of going from the physical illusory reality to the universal view of reality. We humans have the option to discover the deeper truths about our existence and reality.

Rethinking all our perceptions on the nature of reality certainly would not have been necessary if we enjoyed universal harmony, health, and goodwill. This would not be necessary if our personal lives were not threatened by the dangers, strife, and turmoil of daily existence. Something out there just is not working for us. Unfortunately, we humans perceive reality through the lens of our own senses. Herein lies precisely the cause for our dilemmas inasmuch as our five senses provide a five percent quality of human perception. We must begin to question these scientific notions concerning our lives and the nature of the universe.

The original self, the cumulative effect and cassette soul-consciousness, is the core structure out of which the entire multiple system develops. Our behavior, decisions, reactions to our environments, our fears, and moments of sustaining enjoyment evolve directly from the results of cumulative lifetimes.

The issue of the survival of soul-consciousness after physical death is a question that has been raised in every age. Probing the physical aspects of the universe will not furnish the answer to our question. The physical that comprises only one percent of any reality, and the five senses, which contribute an additional five to seven percent of our understanding of reality, must finally be dismissed as a basis for proving or disproving existence after the death of our physical matter-bound bodies. I am quite aware that this appeal to you appears repeatedly in all of my writings. Nevertheless, I have found this to be the one stumbling block for mankind in achieving an information revolution. Our scientists, for whatever reason, hesitate to drive home the accepted doctrine of the uncertainty principle. Physicians and researchers in the field of mental health continue to rely on the western scientific view that matter is primary and consciousness is a property of complex material patterns.

Even those who view consciousness as non-material and formless consider non-ordinary modes of consciousness as something associated with the Divine. These non-ordinary experiences are referred to as trans-personal because they make contact with a reality that goes far beyond the present scientific framework. We cannot, therefore, expect scientists to contradict or confirm this view of consciousness, since at this present stage, they are ill-prepared to fully understand that which is beyond their own five percent of consciousness.

The nature of consciousness is a fundamental existential question that confuses, while at the same time, fascinates those professionals whose livelihoods involve problems relating to it. In spite of all their books, articles, and discussions, the lay-person with a problem is left to fend for himself. How, in good judgment, can a physician prescribe remedies or solutions to mental problems when the practitioner's access to his own brain power is said to be five percent? But, as I have mentioned on numerous occasions, the Age of Aquarius will usher in a people's knowledge revolution. As it is written, "They shall teach no more every man his neighbor or every man his brother, saying, know the

Lord [the Lightforce], for they shall all know Me from the smallest to the greatest of them."[62]

Many of the challenging topics discussed in all of the sciences today, especially the nature of the mind and human behavior, present science with a crisis unparalleled in history. Nonetheless, scientists stubbornly cling to their egocentric positions despite the acceptance of an uncertainty principle that they themselves created. For centuries now, we have assumed that however abstruse an aspect of nature may have appeared, science would always find the answer.

Only in the past few decades has the scientific community come to realize that we are faced with a bewildering and confusing array of complex environmental and life forms, which present challenging problems. Unfortunately, the present scientific establishment is becoming increasingly fossilized by its own particular world view. One cannot continue to create formulas and at the same time inject into them the ever increasing aspect of uncertainty.

Our individual ego is the main cause that fragments us from our true being, namely the Lightforce. Take, for example, the scientific empire, which is founded and based on research that the scientist himself has pre-arranged. Therefore, we cannot, as lay people, differ with their ideas because we are not the true authority. We are told how to cure our bodily ills, how our universe functions, and the probabilities of Life-forces other than our own. There are so many absolute truths given to us through the eyes of people with tunnel vision that it is indeed difficult to imagine how and why we take their knowledge as absolute fact. How wonderful it will be in this Aquarian Age, when knowledge will be the domain of all people rather than a select few.

The Kabbalah, itself, has been for too long a jealously guarded secret. But the time has finally come for it to reach the masses with its message of simplicity. Because in the final analysis, knowledge understood by the layman is to be considered true knowledge.

> *In your compendium, Rabbi Shimon Bar Yochai, the Zohar, the Book of Splendor, shall Israel and the world in the future taste from the Tree of Life, which is the Book of Splendor. And the world shall go forth from its exile with mercy.*[63]

The future the *Zohar* refers to is here and now. As it is stated in the *Zohar*:[64]

> *Alas for the world when Rabbi Shimon shall depart, and the fountains of wisdom will be closed, and the world will seek wisdom, but there will be none to impart it. The Bible will be interpreted erroneously because there will be none who is acquainted with Wisdom. Said Rabbi Yehuda, "The Lord will one day reveal the hidden mysteries of the Torah, namely at the time of the Messiah, because 'the Earth shall be full of the knowledge of the Lord, as the waters cover the sea.'"*[65]

The route to the new physics of the future lies beyond the dimensions of the physical reality of our world. It will permit us to go beyond space-time in our analysis, and hopefully, one day a door will open, "no wider than the eye of a needle, and unto us shall open the Supernal Gates"[66] exposing the glittering interrelatedness of the universe with all its beauty and simplicity.

A commitment to self-knowledge and self-improvement is the first requirement of any individual who wishes to take control of his life and alter his destiny, if necessary. Once that commitment is made, the results can be immediate and fulfilling. Not only will we be happier in the quest to elevate our soul, but we shall also find that pursuit of this goal begins to alleviate a great deal of the suffering initially dictated by our *tikkun* pattern.

The science of the Kabbalah does indeed answer many of the enigmatic aspects of nature, yet it still remains simple. The Kabbalistic vision of reality is based on an in-depth perception of the Bible's coded narration and tales. The Kabbalah teachings provide the underlying laws and principles by which we can establish power and influence over our environment, both terrestrial and extraterrestrial, and understand the power that nature holds over us.

Before concluding our chapter on the mind-brain body connection, let us explore some of the descriptive incarnations provided by the Ari.

Following the sin of Adam, the multiple souls of Adam were incarnated in the generation of the Deluge.[67] Consequently, these same individuals were corrupted with the identical cassette of their prior incarnation. They, likewise, did not complete their tikkun. This is indicated by the verse, "And it repented the Lord that He created Adam (Man) on the Earth and it grieved him at his heart."[68] They were the actual children of Adam.

They were then incarnated into the generation of the Tower of Babel.[69] This is evidenced in scriptures when it states, "And the Lord came down to see the city and the tower that the children of Adam (Man) built."[70] The Zohar and Midrash explain the words "the children of Man" as meaning literally the children of Adam again incarnated.[71] Here again, within their present second incarnation, their life course was prescribed by their actions inherited from the Adam sin experience.

They were then, for the third time incarnated into the evil generation of Sodom,[72] indicated by the verse "and the people of Sodom were wicked."[73]

Following these three incarnations, they were then incarnated for a fourth time in Egypt, as the nation of Israel, and then they began to move towards their tikkun.[74]

The law of *tikkun* is really the law of fair play. By permitting a soul to sojourn in the physical world, the soul is given an opportunity to correct misdeeds performed in a previous lifetime. It is unfortunate that it usually takes us far more lifetimes to complete a *tikkun* than might be required if we would only grasp the problem and apply ourselves to it instead of dwelling in unhappiness over some imagined injustice.

Usually, these lessons are patiently repeated day after day, year after year—even lifetime after lifetime—until the knowledge we have ignored comes crashing in on us, sometimes in a most devastating fashion. And even then, many and possibly the majority of us, do not pick up on the experience and make the necessary *tikkun*. The world, as far back as recorded history permits, presents a society of mankind knowing and learning almost nothing. For example, we still lift our hands against our fellow human being without realizing that warfare does not spare the victor. He, too, becomes the victim of his own insolence and inhuman behavior.

A majority of the Jewish people have least understood this and are therefore most ill-prepared to consider the cassette-forming track of the unfolding of their lives.[75]

In the year 1492, a tragedy for the Jewish people unfolded in Spain where Queen Isabella and King Ferdinand issued a decree of expulsion that sealed the fate of the Jews in that country. It was decreed that, within four months, all those Jews who refused to renounce their faith would be compelled to leave Spain. One hundred years before, in that fateful year of 1391, the Jew in Spain, impoverished, crushed in morale, enfeebled in numbers by conversions forced at the point of death, decimated by massacres, and already on the verge of final dissolution, still did not learn from recorded history the re-run of his pre-recorded cassette. No, the Jewish people would continue a struggle to the last without learning about their collective *tikkun*.

If we could only learn to cooperate with the strings of the universe and the steadfast onward movement of evolution, instead of stubbornly resisting it, our spiritual growth would blossom. Then, and only then, could we achieve the paradise beyond the horizon we all so intensely crave. It also is unfortunate that so few of us wish to take advantage of the memory of experiences through which we have gone. In those experiences lies all wisdom, the reasons for our existence and our educational tools. Yet many of us are hesitant to delve into our own natures for fear of what we may find there.

With the study of Kabbalah, unfounded fears and uncertainty gradually disappear to the point where we begin to sense a kind of control over our actions, destiny, and more importantly, over the hostile environment that we consider home. Instead of experiencing constant confusion in our stream of consciousness, the study of Kabbalah restructures our mental computer so that events in the inner world begin to come forth in a very certain, orderly, and much more quantum-like manner. The study of Kabbalah permits us to leave the driving, the headaches, the ups and downs, and the uncertainties to a consciousness that does not really have any deterministic influence as to how things will ultimately work out.

Our efforts should be blessed with overall objectives that will benefit both ourselves and mankind. How the physical, illusionary world reacts to crises and uncertainties should not be of any concern to us once we have taken a firm position with regard to our singular free choice possibility, namely, Restriction.[76] Humanity's finite aspect, which might be described as the flesh and bones, is subject to the fragmented, crisis-ridden Cartesian rules and regulations. Our 99 percent reality operates beyond limited physical jurisdiction. Only the finite is subject to pain. The Infinite is part of the eternal.

Although the finite, physical reality is rooted in the physical world of limitation, the finite is free to connect and merge at will with the Infinite, which is characterized by certainty, happiness, and freedom from trauma. Connecting with the Lightforce requires paying constant homage to the original act of Creation, which was Restriction. Doing this will assure our physical and mental well-being.

Chapter Three

HUMAN MYOPIA

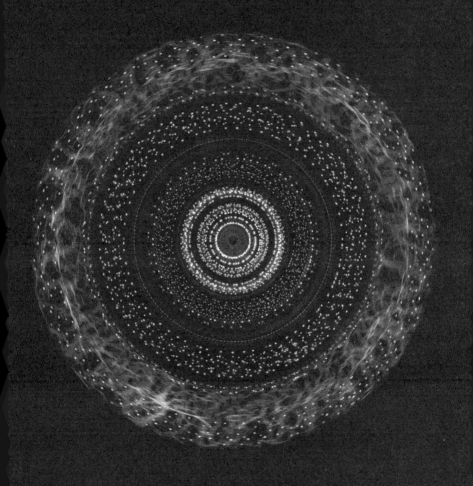

Chapter Three

HUMAN MYOPIA

IN THE BOARDROOM OF A LARGE MULTINATIONAL PLASTICS manufacturer, a division director of sales clicks his way through a series of exciting slides. Each one represents an exotic portrayal of future merchandise packaging and food containers, all made of plastic. Now, all of mankind's food requirements will no longer be packaged in heavy, bulky paper receptacles or metal cans. New, lighter-than-air trays, bags and bowls will provide an assortment of foods from TV dinners to baby foods.

One day, the question will be asked, "Are any of these new plastics biodegradable?" Today, however, for convenience and expediency, questions pertaining to our future welfare and the welfare of this planet are rarely taken into consideration. However, environmental protection agencies all around the world do concern themselves with protecting the environment. These agencies warn us that plastics and nuclear waste do create a serious universal problem. Given the desperate state of our environment, it seems clear that we no longer have the luxury of tunnel or myopic vision.

We no longer can afford to ignore the long term effects that new technologies might have on our planet. The price of our shortsightedness now compels us to live in a constant state of crisis management. The technologies we have created are severely disrupting and upsetting the ecological processes that sustain our planet's natural environment. The poisoning of our water supplies and the contamination of the very air we breathe by toxic chemical waste pose the most serious threats to our very basic existence.

The enormous amounts of hazardous chemical waste are the direct result of the effects of advanced technology and economic growth myopia. The chemical companies have made one attempt after another to conceal the dangers of their manufacturing and their resulting waste hazards. Nevertheless we are becoming an informed people, the benefit of the Aquarian Age. And while we know that serious accidents are occurring, industry is, nonetheless, pressuring politicians to avoid full inquiry. Now we realize that accidents have occurred and that politicians have been pressured by industry to minimize or ignore them.

The entire fabric of life that took generations to evolve is rapidly disappearing. When acid rain falls on rivers, lakes, and oceans, it is absorbed by fish, plants, and other forms of life, thereby polluting the entire ecosystem. In addition, the health hazards brought on by the release of radioactive substances that affect all living organisms are gradual but they do exist. The effects may not be felt for many years and may only appear in future generations. In fact, when we consider the health hazards of the radioactivity of substances released by the nuclear industry, we realize that there is no safe level of radiation. I know that this statement is contrary to what the nuclear industry leads the average person to believe.

Even now, medical scientists generally agree that there is no evidence of a level below which radiation is harmless. Who knows what amounts will produce mutations and disease? It is no wonder, therefore, that medical problems and diseases are reported to be on the increase despite the claimed progress of medical research.

Many nuclear accidents have already dotted much of our planet. Major catastrophes, such as Chernobyl, have occurred or have often been narrowly avoided. We are sitting on a time-bomb, waiting for it to go off at any moment. Who will remain the lucky ones? The effects of a nuclear accident are similar to those of an atomic bomb. We turn again to that major problem upon which myopia due to greed causes acceleration of nuclear accidents and we ask, "How in the future are we to dispose of nuclear waste?" Vital information, such as the fact that each nuclear reactor annually produces tons of radioactive waste that remains toxic for thousands of years, must be released to the public. People must know that no permanent, safe method for disposal has yet been found.

As I have often mentioned, the people or businesses responsible for this grotesque irresponsibility feel no remorse or pain when they dump thousands of dangerous chemical compounds into the ground, rivers, and streams. Human myopia prevents these irresponsible people from considering the problems created for future generations as well as for their own offspring. They have succeeded in making us believe in the great benefits of nuclear weaponry, reactors, chemical food additives, synthetic fibers, plastics, and pesticides.

From a kabbalistic world view, the problem with their reasoning is as follows: If indeed their greed creates human myopia, then it is *not only* future generations that have been placed in jeopardy, *but also* they, themselves, who will now become the victims of their own irresponsible actions. We must constantly be aware that quantum mechanics dictates that our universe does not exist in a fragmented state but has a unified direction. Evil actions are immediately followed by consequences affecting the perpetrator. These perpetrators do not get off scot-free. They suffer the consequences of their actions unless a correction is made in this or future lifetimes.

This statement, however, is no consolation for an environment that will soon not permit us to breathe air, and will contain only poisoned foods and water unfit for human consumption. Because the production and consumption of these food poisons, alien to the human organism, continue to dominate, many countries around the world are grappling with a crisis in increased illness, disease, an enormous dependency on alcohol and drugs, and hospital care.

Chemical manufacturers convince us in their advertising slogans that our life has become enhanced thanks to the use of pesticides for plant life and steroid injections for fowl and cattle. They tell us that hunger threatening the world's population is being reduced. As pesticides destroy the insects that threaten our food production, there will be more food for everyone. Whereas previously, these menacing worms would infiltrate and destroy our fruit, now pesticides can exterminate them. The claim is that the beneficiaries of the pesticide programs include the farmer, the chemical manufacturer, and of course, you and I, the consumer. With more apples now on the market, the prices for apples and other fruits can be kept to a minimum.

Once again, because of human myopia, which is just another expression for the Desire to Receive for Oneself Alone, we are blind to the obvious consequences of these *benefits*. The farmer assumes he is an immediate beneficiary because he is increasing his production and reaping greater profits. The problem, however, is that his fruits now contain cancer-causing agents. Furthermore, the farmer, chemical manufacturer, and consumer may not be reaping the benefits they all believed were inevitable.

More and more people are realizing that the chemical industry has become life-destroying instead of life-sustaining. Loss of life, decreased human productivity, and medical expenses far exceed any of the benefits we may have dreamt of as a result of chemical technology. For example, while some farmers were able to triple their yields per acre and cut their labor force, most of the traditional single family farmers were forced to leave their land.

The long term effect of chemical dependency in farming has not only proven disastrous for the health of the soil but it has also reared a new generation of cancer-ridden young people. Many of these health hazards are further aggravated by the fact that our health care system is totally inadequate to deal effectively with this crises. Current health care is no longer directed by the "physician healing the sick" but rather by business-oriented administrators. This is not to say that health care should not be properly administered. However, the forces administering health services are all too often driven by corporate values that turn health care into a commodity to be sold; one that does not satisfy the health needs of the patient.

Our current health care system favors an approach of profitability for corporate industry that is becoming very expensive and unhealthy for the patient. In the final analysis, the farmer, the health industry, and the chemical establishment will have created a climate where only government can intervene and cover the costs. But where is government to receive the necessary funds if not by making every one of us pay for it all?

The kabbalistic world view, along with its teachings and doctrines, can hopefully induce a people's knowledge revolution, which will force and create change, a change nothing short of a cultural revolution. I am taking this bleak and dismal outlook on the state of our health and physical well-being for two reasons. Firstly, we must come face to face with what appears inevitable. We now stand at the brink of a complete global breakdown, which has resulted in the creation of an unsafe environment. Health hazards, along with medical problems, have reached such proportions that neither health organization schemes nor government intervention can result in the creation of a healthy society.

Secondly, we must acknowledge that despite these ominous warnings about our future health and unsafe environments, there is, nevertheless, a state of balance within the cosmos. Illness and negative human behavior, from a kabbalistic perspective, are not thought of as solely an isolated process. When we tap into the negativity that exists in the cosmos, the result is illness and disharmony. The nature of all things is seen as either in balance and in harmony with the cosmos or in a state of disharmony with the fundamental principles of a dynamic cosmic reality.

To apply the kabbalistic model as a way to develop and understand a quantum *wholistic* approach to universal interconnectedness and interdependence, we must deal with two questions: [Please note that our spelling of wholistic is intentional. We are looking for a total or whole approach in our life patterns.] To what extent is the kabbalistic view wholistic? Which of its aspects can be adapted to our cultural environment and surroundings?"

Simply stated, from a kabbalistic point of view, all things lead in one direction: "Quantum," the substance of which is "Love Thy Neighbor." When this is achieved by mankind, the entire universe, both the seen and the unseen, will be revealed as it actually is, namely a single unified whole. Our universe is perceived as fragmented only because mankind is fragmented.

In the 1940s, man initiated the splitting of the atom which was then hailed as a scientific breakthrough. Only now has it become evident that the use of nuclear power as an energy source is a bad mistake. Nuclear energy, which requires the splitting of the atom, represents the most extreme and dangerous case of technology out of control. Let us therefore re-examine this once highly acclaimed solution for mankind's needs.

The first thought in separating the atom was the idea of creating energy far more powerful than anything that ever before existed. When Einstein discovered his Theory of Relativity in 1905, he made this dream a possibility.

The atom is a beautifully structured entity that corresponds to the balance and harmony of our universe. It is the building block of everything and anything with which we are familiar. When science *decided* to split the atom, mankind then entered into the final threshold of our Aquarian Age. We are forcibly reminded of the passage in the *Zohar*[77] that discusses the two ways in which the final stages of the Aquarian Age will become a reality.

> *Rabbi Shimon Bar Yochai lifted up his hands, wept and said, "Alas for him who will live at that time [the Age of Aquarius] and happy for those who will live at that time ... a pillar of fire will be suspended from Heaven to Earth for forty days, visible to all nations. Then Messiah will arise from the Garden of Eden ... and he will become revealed in the land of Galilee. On the day, the whole world shall be shaken and all the children of man shall seek refuge in caves and rocky places. Concerning that time, it is written, "And they shall go into the holes of the rocks, and in the caves of the Earth, for fear of the Force, and for the glory [manifestation] of His majesty [Messiah], when He arises to shake the earth terribly."*[78]

The Messiah will arise ... and gird himself with weapons of war on which are inscribed the letters of the Tetragrammaton.

Then Rabbi Shimon wept, and the disciples also. Said Rabbi Shimon: "Behold, I was moved a while ago to meditate on the mystery of the letters of the Tetragrammaton, these letters gave me the power of the Lord's compassion. For the mystery of the Lightforce's compassion over His children are imbued within the energy of these four letters. It is now fitting that I should reveal unto this generation something that no other has been permitted to reveal. For the merit of this generation sustains the world until the Messiah shall appear." He then bade his son Rabbi Elazar and Rabbi Aba to stand up, and they did so. Rabbi Shimon wept a second time and said, "Alas, who will endure to hear what I foresee. The exile will drag on. Who shall be able to bear it?"

Then Rabbi Shimon rose and spoke, "Lord our God; other forces besides You have had dominion over us. But By Thee (BeCha=22) only do we make mention of Your Name."[79]

The Name "By Thee" (BeCha) symbolizes the Name of God comprising twenty-two letters.[80]

At that period when there is perfection, peace, and harmony, the two Names are not separated one from another. It is forbidden to separate them even in thought and imagination. But now in exile, we do separate them. "Apart for You," being far away from You and being ruled by other powers ... Israel was involved in many wars "until the darkness covered the Earth." These are the veiled mysteries.

What seems to emerge from the *Zohar* is that through man's knowledge and intervention, mankind has the ability to create fragmentation in the universe. The result of this human activity is the awful pillar of fire and devastation. If we wish to avoid this terrible fate, we must come to the realization that what takes place on this mundane physical level, including the so-called western advances in atom fragmentation, are only reflections of our advances in *inhumanity.*

Fragmented atoms can spell only darkness, as the *Zohar* indicates. The disintegration of the atom began because of our ongoing inhuman actions towards our fellowman. The bombing of Hiroshima and Nagasaki was only the first result. The long-term effects of nuclear power are still to be felt.

In other words, fragmentation, in whatever form it appears, belongs to the dark side that shall "cover the Earth." There are no solutions to the devastation caused by a fragmented atom. However, let us make no mistake about the following point. When it comes to the development of man's most self-destructive system, from which there is no escape, the scientists involved were in a state of robotic consciousness. They were merely reflecting mankind's destructive inclination and activities.

For those of us seeking to enhance the welfare and well-being of humanity, Kabbalah techniques can prove to be a method by which we can avoid the inevitable pitfalls that lie ahead of us. These techniques can and will transform undrinkable water to a quality whereby we will not suffer or undergo the negative results of contamination. This feat is recalled in the Bible, where the Egyptian water was filled with blood but in Gershom, where the Israelites lived, the water was pure.

I know that to many readers of kabbalistic teachings this statement might border on the edge of quackery or appear strongly suspect. However, as I stated in my previous writings, these statements, as dramatic and startling as they may appear, are the recorded teachings of the Masters of Kabbalah. The points raised in the teachings of Kabbalah have been confirmed as going directly to the heart of any subject matter that science has been capable of reaching.

To understand human nature, we study not only its physical and psychological dimensions but also its metaphysical implications. We require an enormous concentration of consciousness to sift through the maze of unnecessary and sometimes even misleading information. In our own day, we are completely inundated with so much irrelevant information that the task of separating the right from the wrong, the important from the unimportant becomes too overwhelming. Consultants, advisers, you name them, are always being called in to assist in sorting out the good from the bad.

Stretching this concept to its limits, the *Zohar* considers the possibility of segregating the good from the bad even when it relates to inanimate life. Just as we comprehend why we separate the good from the bad, a chore we are faced with on a daily basis, so too we can begin to broaden our awareness concerning our ability to act upon physical material nature.

The Age of Aquarius has ushered in new phenomena, unthinkable only a century ago. The separation and fragmentation of the atom followed an era where the idea of an atom even existing as a force was unknown to science. Atomic energy obviously requires an atom. No such creature was born into physics until the beginning of the 20th century. However, the atom as an idea—as an invisible substratum of an elemental substance below the world of physical manifestation—is already mentioned in the *Talmud*. In the 18th century Newton wrote that:

> *"…the Lord, at the outset, created matter in solid, hard, impenetrable, moving particles of such sizes and figures, and with other properties, and in such proportion to space, as most conducive to the end to which He formed them."*

James Clerk Maxwell, a famed Scottish physicist, was equally dedicated to the Newtonian idea of a hard, impregnable, mechanical atom. In 1873, in *A Treatise on Electricity and Magnetism*, he wrote:

> *"Though in the course of ages catastrophes have occurred and may yet occur in the heavens, though ancient systems may be dissolved and new systems evolved out of their ruins, the atoms out of which*

the sun and other celestial bodies are built—the foundation stones of the material universe—remain unbroken and unworn."

At the turn of the century, the German physicist Max Planck was sure that if atoms existed they could not be purely mechanical. He believed that the outside world was something out there, independent from man, something absolute and fixed. Furthermore, he and those who followed him could only view our universe as always passing with some change and thus slowly running down to randomness and ultimately resulting in disorder and rotting away.

Quantum changed all this, but it left the physicist holding a bag full of surprises, asking more questions than ever before. The prior vision of a universe running only one way, towards death and decay, changed with the new age of physics, which allowed the universe to run equally backward or forward. Thus the butterfly could turn back into a caterpillar and an old man into a child. Unfortunately, this mechanistic teaching provides no explanation for the fact that these things do not seem to happen. The old man does not seem to revert back to a child.

However, the reincarnation doctrine states that the old man or woman does again turn into a child. The illusionary period from death to rebirth belongs to the uncertain, illusionary frame of reference to which most of mankind belongs. This might be compared to a passenger on a subway train getting on at one station, traveling through darkness to arrive at the next one. Simply because the train is unnoticed between stations does not in any way alter the system or route of its destination. The fact remains that it is the same train except for an illusionary absence between stations.

This is also true for a soul traveling along the *tikkun* route of destiny. At death, the soul, as the train, seems to disappear for a short while between stops. However, the soul, as the train, makes its appearance again. The illusionary body appears to rot away. However, on the reality level, both forwards and backwards, order and eternity rule supreme.

Science has yet to achieve an awareness of this reality level, and, therefore, it flounders in the abyss of darkness. This uncertainty was best expressed as early as 1894, when Robert Cecil, Chancellor of Oxford University and former Prime Minister of England, catalogued the unfinished business of science with regard to atoms in *Evolution: A Retrospect. The revised address delivered before the British Association for Advanced Science, Oxford, 1984, p. 27:*

> *"What the atom of each element is, whether it is a movement, or a thing, or a vortex, or a point having inertia, whether there is any limit to its divisibility, and, if so, how that limit is imposed, whether the long list of elements is final, or whether any of them have any common origin, all these questions remain surrounded by a darkness as profound as ever."*

Just a mere century ago, there was little known about atoms. However now, the atom not only dominates the existence of our very lives but also may well be the cause of planet Earth's demise. Science has developed unconscionably and has essentially no moral or ethical guidelines. What oath of allegiance do scientists impose upon themselves? Their results can be far more devastating than those of the medical profession. There is, after all, the Hippocratic Oath, which injects some form of responsibility into the medical profession.

The influence of business on all the sciences has brought about a disturbing imbalance within every facet of human and ecological existence. Only now have the effects of business become major problems. Just as the petrochemical industry has convinced the farming industry that Mother Earth, our soil, needs massive invasions of chemicals, the pharmaceutical corporations have convinced the medical establishment and its patients that to achieve good health, the body requires continual drug treatments and medical supervision.

I did not write this book to deride business, the science community or the medical profession. My intention here is twofold: Firstly, I want to raise the consciousness of a sleeping mankind to an awareness of what is going on within our midst. If we are to believe the statistics of the cancer, heart, arthritis, and other foundations, it appears that most North Americans do not enjoy perfect or even satisfactory health. And if there are some who still claim things are just fine, then we are reminded each day of the threatening drug epidemic that no longer permits us to live safely in our own homes or walk our streets free of worry from burglars or muggers.

Secondly, I want to make mankind aware of a time honored system, a two thousand year old one, for that matter, that

Rabbi Shimon Bar Yochai introduced in his *Zohar*, and is more fully elaborated upon in Rabbi Isaac Luria's *Gates of Meditation* and *Gate of the Holy Spirit*. These works provide us with the tools to obtain total and overall control of our inner space when all around us is crumbling, when all around us everyone searches for understanding.

Richard Feynman, a noted theoretical physicist, when addressing undergraduates at the California Institute of Technology, raised this question, "What do we mean by understanding something?" His grasp of human limitations, plus the recognition that we essentially use only five percent of our mind while the rest of our mind is asleep, forms his answer in his book *Six Easy Pieces: Essentials of Physics*, 1995:

> *What do we mean by "understanding" something? We can imagine that this complicated array of moving things, which constitutes "the world" is something like a great chess game being played by the gods, and we are observers of the game. We do not know what the rules of the game are; all we are allowed to do is to watch the playing. Of course, if we watch long enough, we may eventually catch on to a few of the rules. The rules of the game are what we mean by fundamental physics. Even if we knew every rule, however, we might not be able to understand why a particular move is made in the game, merely because it is too complicated and our minds are limited. If you play chess you must know that it is easy to learn all the rules, and yet it is often very hard to select the best move or to understand why a player*

moves as he does. So it is in nature, only much more so; but we may be able at least to find all the rules. Actually, we do not have all the rules now. (Every once in a while something like castling is going on that we still do not understand.) Aside from not knowing all of the rules, what we really can explain in terms of those rules is very limited, because almost all situations are so enormously complicated that we cannot follow the plays of the game using the rules, much less tell what is going to happen next. We must, therefore, limit ourselves to the more basic question of the rules of the game. If we know the rules, we consider that we "understand" the world.

The idea presented by this famous physicist is that true understanding essentially is something that appears as a highly unlikely achievement. There is just too much out there and their interconnectedness makes matters all the more complicated. Then what hope does the future have in store for us when everything we relate to, or participate in, contains so much uncertainty?

I see in the uncertainty principle, as presented by science, the grandest design of the Aquarian Age. Behind the profound structure of Newtonian classical physics, that in only three centuries had begun to reshape the entire human consciousness, lies a basic commitment to a mechanistic view of life. The trend was away from religious doctrine, which had dominated people's lives for so long. This new civilization was deeply committed to beliefs of nature that were totally different from those of earlier civilizations.

This Newtonian perception, which no longer dominates scientific thinking, was too rigid. What was necessary therefore, was a complete transformation to the new age of physics. However, now again more change is necessary. We must act positively and assuredly for we stand on the brink of destruction for all mankind. The illusionary reality must be replaced by true values. If we continue to pollute our waters for profit we will soon not be able to have clear drinking water. The illusionary profit seen today, will be but the destruction seen tomorrow. When we destroy each other in business for fast profit, the end result must be faulty airplanes, acid rains, and war. Kabbalah and its doctrines are now ready to replace uncertainty and illusion for the reality and wholeness of a true eternally providential nature.

Kabbalistic teachings provide the individual with the ability to create an internal, private inner space universe, whereby all the routines of uncertainty, chaos, disorder, and destruction are considered as ineffectual illusionary states. Sounds too good to be true? And yet, the Bible is replete with descriptions and narrations of this Aquarian Age phenomenon.

Becoming a student of the Kabbalah requires a profound commitment to the kabbalistic world view. The first step in this commitment is the realization that Satan, our own particular ego, is the force that threatens our very existence. Ego is the underlying factor for the limited expression of our five percent consciousness. When all around us has become uncertainty, the ego prepares us with a convenient memory lapse. Our ego convinces us that all our decisions and activities are the direct result of our own conscious mind and thought.

As corporate managers, we make decisions that are detrimental to the welfare of consumers, patients, and others, as well as subsequent generations and offspring. As business decision makers, we become nearsighted and only encompass our present egotistical positions and immediate rewards. As a result, the total environment and nature as a whole will suffer greatly for our inexcusable lack of Desire to Share on a quantum level.

The next step in the commitment process requires an open-mindedness towards any and all of the information presented by kabbalistic teachings. At first, this might appear to be a simple matter. But considering the built-in programming that most of us have undergone during our early years, the unshackling of our preconceived ideas is the most important single factor preventing the people's information revolution. To overcome this obstacle will require a monumental effort and commitment if we are to gain control over our polluted environment.

The fragmentation in every facet of our lives has wrought upon us a nature that is profoundly inhumane. Large corporations formulate the way we live and behave. Exploitation of our natural resources, unfair competition, and resorting to intimidation or bribery are often aspects of today's corporate activities. The maximizing of profits, which by and within itself is commendable, becomes the ultimate good, to the exclusion of all other considerations. To *succeed*, corporate executives must leave behind the humanity they, hopefully, employ in their personal lives. When they enter the halls of big business, there is no room for feelings or regrets.

Whether by dire need or spiritual enlightenment, the inevitable revision of our basic concepts and theories must be so radical that the question arises as to whether or not our present system will survive it. From a kabbalistic world view, current frameworks are outmoded. We must again address ourselves to the significance of human attitudes, lifestyles, and real values. We must deal with human potentialities and integrate them into the underlying matrix of our entire global system. The kabbalist sees, in this ultimate system, a side-by-side working arrangement between the spiritual and the scientific.

For those individuals prepared to accept the true and obvious realities of life, let us now proceed to the kabbalistic perspective of the dilemmas that confront us in the Age of Aquarius. How will we overcome their impending disasters?

The primacy of physical appearance is a fact of everyday life, which neither the scientist nor the kabbalist can ever escape. Both of them must always return from their laboratories or meditative states of consciousness to what is commonly referred to as the world of *botz* (mud). For the present, this physical mundane appearance maintains its grip by almost never being the least bit changed by discoveries drawn from science or the Kabbalah. The Surgeon General, armed with laboratory tests, warns against smoking, and yet millions continue to smoke. The kabbalist returns from his meditation to offer the doctrine of Restriction to create positive effects for the individual and the world, and yet millions continue to ignore his message.

The world still continues on its merry way towards polluting the food we eat, the water we drink, and the air we breathe. Why? Because we cannot discern or differentiate between that which is real and that which is illusion. The appearance remains the same. The air we breathe has not taken on another appearance. The food we consume daily looks even better and shinier. The packaging has even been improved. Water for the most part retains its same visible color and taste. Despite laboratory analysis, nothing has really changed.

The relentless search by modern science for the foundations beneath mere outward appearances has given new validity to the belief that physical appearances may well be a contrived illusion. We are now beginning to "read between the lines." But there is nothing more between the lines than what we can observe. We simply make use of expressions that we do not examine seriously enough. Nevertheless, science has finally reached a level of awakening. Even scientists agree that we cannot consider the physical reality as anything more than illusionary.

The results of science have been rather perplexing, if not downright confusing. Scientists cannot give a full account of "reality" in accessible language that can be scientifically demonstrated in the laboratory, nor can they test reality effectively through test-tube technology. It seems that manifest physical existence must dominate our thinking and behavior, to the exclusion of the underlying non-physical reality and causation. The fundamental question is, "How shall we exist in the two worlds, the physical and metaphysical?" We must always be cognizant that what we perceive on our mundane level should be in conformity with the Desire to Receive for the Sake of Sharing. We must constantly be aware that our daily actions are in harmony with the aspects of sharing.

On the surface, in everyday life, things often appear to be without arrangement or form. However, on closer examination, there seems to be a hidden order. Let us take, for example, Morse code messages. To one unfamiliar with the code, it is a seemingly random jumble of sounds but to the initiate, the code is an intelligent signal. When information is hidden in something else, it can be extremely difficult to discover.

These two parallel universes have been identified in terms of the Tree of Life existence of reality, the true ordered level of consciousness, and the Tree of Knowledge of Good and Evil phase of our illusionary reality. It is here that randomness, uncertainty, chaos, rot, disorder, illness, and misfortune make their presence felt. This uncertainty holds that the farther we look into the internal, true reality of the sub-atomic world, the less clearly we are able to distinguish the features of individual particles. Scientists readily agree that any object, say a chair, does not physically exist until the observer takes notice of its presence. In other words, when I sit on the chair, the chair for me has now become a reality. So for all practical purposes, the chair exists and does not exist, depending on who notices the chair.

This problem has already been dealt with in the *Zohar*.[81] The idea of two realities and how to "be in both places at the same time" has always been known to the kabbalist.

"Say unto Aaron, take thy rod." Why Aaron's rod and not that of Moses? Because Moses' rod was more sacred, energy of the upper Holy Name had been engraved on it, and the Lord did not want it to become defiled by coming into contact with the rods of the Egyptian magicians ... Rabbi Chiya asked Rabbi Yosi: "As the Lord knew that the Egyptian magicians were able to turn their rods into serpents, why did He command Moses and Aaron to perform this sign before Pharaoh? There was nothing spectacular in this to him." Rabbi Yosi replied: "Pharaoh's dominion originated with the energy-intelligence of the serpent and therefore his punishment commenced with the serpent. When the magicians saw Aaron's rod turn into a serpent, they rejoiced because they knew they could do the same. But then Aaron's serpent turned into a dry rod again, as it is written: 'And Aaron's rod swallowed up their rods.'[82] They and Pharaoh were then astonished, realizing that there was a superior power on Earth.

"Thus Aaron showed a double sign, one above and one below. One above by demonstrating to Pharaoh that there was a higher serpent that ruled over theirs. And Below by making the energy-intelligence of wood subdue their serpents.

"Do you think that the magicians referred to were merely producing illusions and make-believe? Their rods actually did 'become serpents,' the rods

physically became manifested as serpents." Said Rabbi Yosi: "Even when their serpents returned to become rods again, the rod of Aaron swallowed them up. As it is written: 'And the rod of Aaron swallowed up their rods.'"

Several startling revelations emerge from the *Zohar,* and when we consider their implications, they should and will provide some of the vital tools we need to meet the crisis of pollution and contamination. The possibility of mind over matter is clearly demonstrated by the ability of both Aaron and the magicians to turn their rods into snakes and then back again to rods. The *Zohar,* additionally, takes the opportunity to make a point of this scenario as something that involves the real reality. Familiar with the idea that "suggestion" and "faster than the eye" techniques might have been involved in the biblical narration, the *Zohar* deciphers the Biblical Code to stress the fact that what is involved in this metaphysical clash is the power of mind over matter.

The *Zoharic* interpretation of the biblical verse is that the rods actually did become serpents, insisting that their accomplishments were not illusions or make believe. We are all familiar with Harry Houdini, the escape artist, and Doug Henning, who it is said, can make the Statue of Liberty and a Boeing 747 disappear before the eyes of ten thousand people. This was not the case in Egypt. Theirs was an awesome power of control over the molecular and atomic structure of matter.

We must always maintain an awareness of our own physical corporeal body, which consists of 99 percent atoms and only one percent of physical matter. Obviously, the dominion of our body rests with the internal, unseen atoms. And what are atoms if not energy-intelligence consisting of protons (positive, Desire to Share energy-intelligence consciousness), electrons (negative, Desire to Receive energy-intelligence consciousness) and neutrons (the centrality, Desire to Receive for the Sake of Sharing energy-intelligence consciousness).[83]

Consciousness is a cosmic scenario. Despite efforts by the scientific establishment to present mankind as having no role to play within the cosmic mechanism, the kabbalistic perspective of man has always been one that places him as the central figure within our cosmos. But mankind today, more than ever before, sees himself as nothing but another component of a massive computer. For many, that computer is perceived as having surpassed the ability of a human being. Man, it seems, cannot and does not alter or redesign the cosmos.

Not so, says the kabbalist. The Bible decreed that man surely could, and would, alter the influence of cosmic order. The Lord recognized the necessity for a cosmos permitting the existence of the mind as a separate entity that can act on matter causing it to behave in apparent violation of the natural laws and principles of the universe.[84]

A further startling revelation, one more fascinating and pertinent to our present environmental dilemma, is the biblical statement that "Aaron's rod swallowed up their rods." The idea that metaphysical energy-intelligences hold sway over all corporeal physical matter is maintained by the *Zohar* when it declares, "One above," by demonstrating to Pharaoh that there was a higher serpent that ruled over his serpent.

When the so-called miracle of Aaron's rod swallowing up the rods of the magicians occurred, the *Zohar* concluded that a sign below, "by making the intelligent energy of wood subdue their serpents," indicated that a system and method of overcoming dark, negative energy-intelligences existed in our universe. The word "serpents" was a code name for consciousness connected with the Dark Lord and related to all negative manifestations that exist for the sole purpose of creating havoc and chaos in our lives. No kingdom could escape the wrath of the Dark Lord when dominion of our environment was passed over to the empire of his death star fleet.

Nuclear waste, polluted air, acid rain, poisoned fruits and vegetables, fish contamination have all come together in our day to create a panic among the people. A ground swell revolution never before dreamt of has overtaken the consciousness of the man on the street. For most people, eating is beginning to seem like a hazardous enterprise, with real risks. The Dark Lord has taken over the very fabric of our lives.

For the present, the prudent move would be to stop breathing. Data collected by the Environmental Protection Agency (EPA) of the United States shows that the air may be far more poisonous than previously thought. Industry routinely releases billions of pounds of toxic substances in the air. From the looks of things, the air should carry a warning from the

Surgeon General stating, "Quit breathing now and greatly reduce serious risks to your health."

Let us pause for a moment and examine the fundamental components that have created the crisis we now have to face. We must explore the origin of pollution and toxic waste before we can begin to arrive at any sensible solutions about how to eliminate these destructive forces from invading our very being.

A good starting point for our investigation is, as always, the cosmic code of the Bible. The *Zohar* declares that the Lord created two basic constellations.[85] An equal cosmic power of good and evil was granted to these two constellations respectively. Thus, the two fundamental systems of good and evil could now exercise cosmic influence over man. The battle between good and evil then began.

The idea of good and evil originated as the Tree of Knowledge of Good and Evil mentioned in the Bible.[86] It is to the sin[87] that Adam and mankind owes its corporeal existence. Born from the pollution of all physical matter by his connection to the Tree of Knowledge, Adam thereby severed the relationship between Man and the outer-space (Tree of Life) connection.[88] Before this, the Upper Reality Realm and the Earth's mundane existence were of one thought and in perfect harmony. The wellsprings and the channels through which everything in the higher Celestial region flowed into the lower realms were still active, complete, and thoroughly compatible. The vessels—all forms of corporeal existence—and the Lightforce were still in perfect tune with each other.

When Adam sinned, the cosmic "thought" connection was severed. The order of things was turned into chaos. The Lightforce was simply too hot to be handled. Raw, naked energy of this intensity was not meant for our world of action.

The outer-space (Tree of Life) thought process was beyond the limits of time, space, and motion. The Tree of Knowledge of Good and Evil realm, with all of its limiting factors, was insufficient to channel the Heavenly communication. Consequently, the Biblical Code continues, "they sewed fig leaves together and made for themselves *hagorot*, (insulated garments),"[89] which was now required to sustain the primal energy-intelligence of the Lightforce. This new development is similar to that of the astronauts who today need specially designed space suits to protect themselves from the perils of outer space.

The doctrine of nakedness now becomes closely aligned with the concept of the outer-space connection. The individual soul retains its own particular existence only in relation to its ability to sustain the energy-intelligence of the Lightforce. Just as live or exposed electrical cables serve no useful purpose, when Adam and Eve were cosmically severed from the thought process by their sinful activity, they were no longer in accord with the dynamic interplay of the all-embracing whole. Their inability to handle the intensity of the Lightforce left them naked.

Thus mankind's polluted, sinful behavior and activity is behind the crisis we face today. The awesome power of the atom is the result of a process in which the uranium nuclei are broken into fragments. The creation of a fragmented atom has led to a profound environmental imbalance generating numerous symptoms of ill health and ill will. The division of the atom is precisely the disrupting scenario created by the sin of Adam.

Every action of man is carried by the channels of the cosmos, whether man knows it or not. Every earthquake, every supernova, every war is the direct result of violence and hatred in the hearts of men. We have, at our fingertips, the ability to recreate Eden. Instead, we build nuclear warheads and prepare for unspeakable hell. Two nations at war with each other spit out their vengeance at one another until they both become exhausted. The cloud of hatred may have vanished but the chaos and the suffering of individual families remains. Mankind should have learned of the futility of war and hatred, but envy and evil eye still remain a part of our human landscape.

Consequently, pollution and contamination, produced by the activities of mankind, can be removed by man's ability to change his ways towards a more positive approach. The only alternative for those who do not hate or envy their fellow man, for those who do not permeate their very essence by their own greed, is through the Light of the Creator, learned through the teachings of Kabbalah.

Where has there ever been a system that can alter cosmic destiny? Despite polluted air or contaminated water, can those positive spiritual people remain unaffected by our chaotic environment? It sounds like a bizarre science fiction story to claim that those people acquainted with the doctrines of the Kabbalah can and will enjoy fresh, uncontaminated water, when people all around them seek fresh water with not a drop to be found. The *Zohar* and Biblical Code do not consider this phenomenon as something beyond the grasp of mankind's thought consciousness. The *Zohar* explains:[90]

> *"And the Lord said to Moses, 'Say to Aaron, take your rod and stretch your hand on the waters of Egypt, on their streams, on their canals, and on their ponds, and on all their pools of water, that they may become blood'"*[91] *Rabbi Yehuda said, "We must concentrate on this passage. How could he have gone to all these places, namely to all the waters of Egypt and all their ponds throughout the land of Egypt?"… He answers, "The waters of Egypt" is the Nile River. All the other ponds and streams and wellsprings and all their waters were filled from there. Therefore, Aaron raised his hand only to smite the Nile. Come and see that it is so, for it is written, "And Egypt could not drink of the water of the River."*[92] *So we see that the River includes all the waters of Egypt.*

What becomes quite evident from the *Zohar* is the quantum effect of Aaron and his spiritual consciousness. Aaron's dominion over the vast expanse of the cosmos was demonstrated when all the waters of Egypt became contaminated by blood. There was no necessity for Aaron to come in direct contact with each of the waters of Egypt.

The *Zohar* continues:

> *Come and see, the Firmament that contains the sun and moon and stars and constellations is the gathering place of the water, for it receives all the water, namely all the lights, and waters the Earth, which is the Lower World, Malchut. As soon as the Earth receives the waters, it spreads them and divides them to every side, and from there everything is watered.*
>
> *During the time when Judgment [negative energy-intelligences] prevail over the world, then the Lower [physical] World, which is Malchut, does not nourish or tap the positive energy-intelligences of the Upper Firmament [of sun and moon], but nurtures from the Left Side that is not included in the Right. Then Malchut is called, "The sword of the Lord is filled with blood."*[93] *Woe to those who then nurture from her and are sustained by her.*[94]

This brief *Zoharic* summary provides us with some idea of the profound difference between the superficial interpretation of the Biblical Code and its comprehension as a cosmic code by the kabbalists. The energy-intelligence of each verse has a bearing upon the dynamic interplay of the universe. If the whole of the universe is to be considered an enormous complex machine, then man is the technician who keeps the wheels turning by providing fuel at the appropriate time.

The energy-intelligence of man's performance and activity essentially supplies this fuel. Consequently man's presence is of central importance since it unfolds against a background of cosmic infinity. The mystical conception of the Bible is fundamental for the understanding of the cosmos along with its laws and principles. The Bible must be seen as a vast *corpus symbolicum* of the whole world. Out of this cosmic code of the reality of Creation, the inexpressible mystery of the celestial realm becomes visible.

The verse in Isaiah, "The sword of the Lord is full of blood," appears to indicate the vulnerability of water and the connection that water has toward the evil side.[95] When man's negativity prevails within the cosmos, water turns into blood or any other form of contamination. In other words, matter or material substance is subject to the dominion of positive or negative energy-intelligences prevalent at any given time.

The three fundamental forces of our universe, the Desire to Receive, Desire to Share, and Restriction (Desire to Receive for the Sake of Sharing), are contained within an atom and are physically expressed and manifested in all material forms. These three forces, designated in physics as the electron, the proton, and the neutron, portray the intrinsic thought energy-intelligences. The electron makes manifest the thought energy-intelligence of receiving or Left Column. The proton expresses the thought energy-intelligence of sharing or Right Column. The mysterious task of neutrons is the creating of a unification of the two opposing forces, proton and electron. In the kabbalistic view of the neutron, its inherent characteristic is the Central Column of Restriction.[96]

The threat of pollution and contamination is the greatest danger facing humanity today. Our present crisis is the direct result of mankind's ability to invade the atomic structure. The preponderance of negative human activity has created a dominion of negativity over the whole of the cosmos.

Consequently, pollution and contamination reflect this imbalance within the universe. Water is influenced and controlled by a positive energy-intelligence factor, the proton. However, with the enormous negative human activity weaving itself through every fabric of our communal life, it is little wonder that the satanic, evil forces of water's negative energy-intelligence now remain in control of our water, food, and global environment.

The situation is apparently hopeless without any relief in sight. Were it not for the kabbalistic knowledge, the future ecology of our civilization might well remain in jeopardy and the end of the Earth might be a foreseeable reality. The kabbalistic truth of the matter is that for those who follow the doctrines of Kabbalah, the future does not appear as bleak as one might imagine. The oft quoted *Zohar*,[97] "Woe unto those who will live at that time [Age of Aquarius], yet happy are those who will live at that time," attests to the fact of a dual cosmic reality. Thus, there appears to exist a cosmos within a cosmos.

The teachings of Kabbalah provide us with the opportunity to segregate ourselves from the polluted physical cosmic reality and connect with the essential unified all-embracing whole. "God," as Einstein remarked, "does not play dice with the universe." He is not a cruel, unsympathetic producer insensitive to the needs of mankind. Despite our overwhelming global frustrations, we can do something about our lives and our world.

As a matter of fact, a similar condition, where the entire world was dominated and controlled by evil forces, is described in the Bible, when it refers to the period of the exodus and Middle Kingdom of Egypt.[98] The idea of two cosmic realities is further elaborated upon by the following *Zohar*:

> *"The sword of the Lord is full of blood."*[99] *Woe unto them who must drink from this cup. At such times the sea takes in from both sides (central-positive and negative) and divides itself into two parts, white from the side of Mercy and red from the side of Judgment. Thus it was the fate of Egypt, namely the red, to be cast into the Nile, and the punishment was inflicted from Above and Below.*
>
> *Therefore, Israel, connected with central-positive energy-intelligences—the white part of Malchut, the white part of the physical illusionary reality—drank water. The Egyptian, drawn towards the negative evil energy-intelligence—the red aspect of Malchut—drank blood.*
>
> *Furthermore, do not understand or interpret the plague of blood to be one merely of a disgusting nature. But come see the significant feature of this plague. When the Egyptians drank the blood, it entered their intestines, until they decided to*

purchase the water of the Israelites, and then they drank water.[100]

Separation of the universe's two realities is strikingly revealed by the *Zoharic* interpretation of Plague One, Blood. The Israelites were in no way affected by the waters turning to blood. In addition, the Israelites' control over water extended beyond their immediate environment. By virtue of the quantum effect, they made certain that water purchased from their fellow Israelites, once purified, remained in a state of un-adulteration even when drunk by the Egyptians. This encouraging thought from the Biblical Code and its deciphering instrument, the Kabbalah, is precisely what will be necessary in our Age of Aquarius.

The summer of 1988, was a revelation of things to come, an experience of the worst of household horrors. The toilet of the ocean was backing up. Medical waste began washing up on the beaches, from Long Island to the New Jersey coastline. In the effort to restore our oceans and farms, as in other struggles for a purer environment, we won the battles but not the war. The moment each small victory is achieved, forces of human greed and negativity are at work against it.

Perhaps these early skirmishes over the environment will escalate to a global consciousness and awareness of the fact that in the final analysis, all of humanity suffers from uncontrollable appetites. Perhaps this contamination and pollution will serve as a model for all people to pull together against the common enemy, namely greed and the Desire to Receive for Oneself Alone. Possibly the shock that greed has caused will mark a turnaround in our regard for humankind and all inhabitants of planet Earth.

Invasion of the marine environment by our persistent dumping of human toxins, radioactive waste, and chemicals is now returning to haunt us. There seems to be no escape other than this reality that within Kabbalah lies Earth's salvation.

There are sections of the *Zohar* called the *Ra'aya Mehemna*, (Beloved Shepherd), the teachings that Moses taught to Rabbi Shimon Bar Yochai. We now quote this from one such section of the *Zohar*:

> *"But they that are wise shall understand,"*[101] *for they are from the side of Binah (Intelligence), which is the Tree of Life.*[102] *And because of these wise people, it is written in the Book of Daniel: "and they who are wise shall shine like the splendor [Zohar] of the Firmaments, and they that are instrumental and responsible for leading many to spiritual righteousness, numbering as the stars for ever and ever."*[103] *Only by virtue of your book, Rabbi Shimon bar Yochai, which the Book of Splendor (Zohar) will Israel taste form the Tree of Life, which is the Book of Splendor (Zohar). Only through the instrument of the Zohar shall mankind be brought forth from exile with compassion.*[104]

The *Zohar* and the *Book of Daniel* are inextricably bound up with each other. Their language is of Aramaic origin. Many legends have grown up around Daniel in Muslim and

Christian tradition as well as in Jewish lore. One tradition supported among kabbalists relates that Daniel and Mordechai returned from Jerusalem to Persia with the teachings of Kabbalah, which subsequently became known as the written word of the *Zohar* by the later Sage, Rabbi Shimon Bar Yochai. The kabbalistic doctrine encoded in the *Book of Esther* provided the Jews of Persia with the metaphysical weaponry by which they overcame the evil, holocaustic decree of Haman against the Jewish people.[105]

The coded *Book of Daniel* is replete with references to the Age of Aquarius. The *Zohar* unravels its mysteries. The secrets that once brought a *temporary* Age of Aquarius to Persia are now ours. In the present day, we await a permanent Age of Aquarius.

For centuries now, we have assumed that however abstruse an aspect of nature may appear, science would always find the answer. Recently, and only in the past few decades, has the scientific community come to the realization that we are faced with a bewildering and confusing array of complex life forms with which the new physics is unable to adequately cope. On the other hand, the deeper we probe, the simpler the task becomes. Soon, we will appear on the threshold of a whole new era of physics.

The route to the new physics of the future, for the layman at least, lies beyond the dimension of the physical reality of our world. The route will permit us to go beyond space-time in our analysis. One day we will open a door no wider than the eye of a needle, and unto us shall open the Supernal Gates exposing the glistening interrelatedness of the universe with all its beauty and simplicity.[106]

Towards the arrival of that date, the *Zohar* holds out more hope than science, which must rely largely on randomness and probability. *The Book of Splendor* can provide a direct link and contact with the universal energy-intelligence and can present the world of metaphysics as an exact, simplified science. The science of the Kabbalah answers many of the enigmatic aspects of nature, yet it remains elegantly simple.

The *Zohar*'s world view of our universe transcends the physical and occupies a frame beyond space-time, whereas the modern age of physics remains fixed and limited to the frames as presented by Einstein. The kabbalistic vision of reality that we have described is based on an in-depth perception of the Bible's coded narrations and tales. The description provided by the Bible sounds quite similar to the description of modern space systems. This description emphasizes the outer space connection as the energy-intelligence systems referred to as the Tree of Life.[107]

> *Rabbi Elazar said: "The Lord will one day re-establish the world and strengthen the spirit of the sons of men so that they prolong their days forever. As it is written: "For as the days of a tree shall be the days of My people."*[108] *This is an allusion to Moses, through whom the Law was given and who bestowed life on men from the Tree of Life. And in truth had Israel not sinned with the Golden Calf,*[109] *they would have been proof*

against death, since the Tree of Life had been brought down to them.[110]

Moses provided a cosmic connection for the children of Israel that would control the decaying factors within the Tree of Knowledge reality, including eternal immortality. Moses drew the all-inclusive positive cosmic energy force by tapping the source of this energy from the Tree of Life.

All the biblical incidents suggest concepts that will take us from our familiar material world of human experience to a more subtle, beautiful view of reality. The kabbalistic teachings are very much a part of the universal thread of our cosmic order, the Tree of Life.

Chapter Four

FISSION OR FUSION

Chapter Four

FISSION OR FUSION

AS I WRITE THIS CHAPTER, THE RACE FOR FUSION, EXPLAINED IN greater detail in the subsequent pages, has broken into a scientific frenzy with the universal hope that we are closer toward caging the power of the universe inside a mundane utility plant. Fusion does not generate greenhouse gases like our present nuclear reactors that warm the planet or raise sea levels and cause droughts. Fusion does not release the gases that have created the acid rain that is ruining Earth's forestry. With fusion, radioactivity is so low that it would not require the necessities of complex waste removal or the burial sites presently needed for today's radioactive waste.

The whole idea of fusion has provoked a great deal of new thought about the scientific approach to energy. Textbooks taught us that there is only one way to persuade atomic nuclei to fuse. That method was brute force. At the very least, fusion has brought us the realization that what is more, might very well be less,[111] and what is less, is in fact, more. Suddenly, the scientist is beaten into the harsh reality that cold might be better than hot.

But I might be running ahead of myself and my readers. What does fission or fusion have to do with the doctrines of Kabbalah? Does the Age of Aquarius have a bearing on these new paradoxical scientific features?

The subject of this chapter is the connection between our physical illusionary reality and the kabbalistic perspective of the world. Physics is solely concerned with the objects and events of inanimate nature. Kabbalah, if it is to be at all satisfactory, must embrace the whole of the physical and metaphysical realities and must deal with questions of thought and the soul, including the problems of human activity.

Max Planck, the early 20th century physicist who discovered the quantum theory, did not believe that the search for an underlying unity in nature leads to mysticism. However, he did concede that scientists have learned that the starting point of their investigations does not lie solely in the perceptions of the senses. "Science," said Planck, "cannot exist without some small portion of metaphysics."[112]

Intangible nature is, after all, part of our universe. Any understanding of the world claiming to be truly total and comprehensive must take into account the laws of intangible nature. Rabbi Shimon Bar Yochai, author of the *Zohar*, strongly addressed this point in the *Zohar* when he asked, "How we can be certain of the *Zohar*'s interpretation of the metaphysical plane?"

> *And for those persons who do not know, yet have a desire to understand, reflect upon that which is revealed and made manifest (in this world) and you shall know that which is concealed, inasmuch as everything (both physical and metaphysical) is the same. For all that the Lord has created in a corporeal way has been patterned after that which is above.*[113]

Thus we learn the sublime *Zoharic* teaching: When the Kabbalah reveals the essence of unseen elements, its interpretation of the concealed will not and cannot conflict with subsequent revealed actions and interactions.[114] The *Zohar* presents us with instant, immediate knowledge of the root of any matter. This seminal knowledge eliminates the necessity of going through the customary procedures of trial and error, action and reaction, and is independent of the fluctuations of time, space, and motion.

This statement now brings us back to the idea of fusion and fission. Neither this book nor the author claim any expertise in the physical sciences. The ideas presented here are, of necessity, simple enough for the layman to understand. While some of the concepts may seem strange or seemingly difficult upon first contact, with some additional effort on the part of the reader, a thorough understanding of the material is within the grasp of all.

Nuclear fission, simply stated, refers to split atoms and is based on a concept involving splitting and fragmentation. Nuclear fusion, on the other hand, is a thermonuclear reaction in which the atomic nuclei of a light element undergo a transformation into those nuclei of a heavier element, with the release of great energy.

Kabbalistically, the lessons to be learned from these two *divergent* paths that supply energy are profound. Because of their metaphysical implications, fusion and fission have escaped the interest of thinking intellectuals and scientists. The primary question that must be raised is, "Why did fission precede fusion?" Furthermore, fusion, with its extreme simplicity, can produce all the electricity the world needs—cheaply, cleanly, and equitably. Developing nations could produce all the energy they needed. Fusion is far safer than conventional atomic power in which the reaction that splits atoms can run amok.

The first message that should emerge from fission experience is that many hazardous by-products are produced. Without even considering the effects of a nuclear war, the damage to our health and environment has been enormous and horrifying beyond our wildest imaginations. Radioactive waste is a situation for which no solution has or will ever be found. The hazards of nuclear reactors themselves, as evidenced by the Chernobyl mishap still vivid in our memory, are a threat to the health and safety of millions of people. Even the smallest amounts of radioactive waste can produce mutations and diseases.

Fission is a process in which uranium nuclei break into fragments, most of which are radioactive substances. As emphasized in the previous chapter, the deepest roots of our current pollution crisis lie in the patterns of human behavior and activity. It has already been stated by the *Zohar* that due to Adam's sin and connection to the Tree of Knowledge, mankind continues to suffer under the heels of a fragmented society. Man's inner fragmentation mirrors his view of the world outside of him as a conglomerate of separate societies and events. This fragmented view can be seen as the essential reason for the present series of environmental and social crises. This fragmentation has brought an ever rising wave of violence, both within the family system and in our society. Our lives have all too often become both mentally and physically dangerous.

Consequently, as long as our view of the world is fragmented, as long as we are under the spell of the Desire to Receive for Oneself Alone, we shall be subject to the hazards of fission or fragmentation. Thus a collective consciousness of fragmentation emerged that directed and forced scientists and physicists to embark upon a path of fragmentary research, which ultimately brought the harsh realities of the nuclear age upon all mankind.

During the 1970s, fundamental physics set out to unify our strange and complex world into a single conceptual framework. Fresh discoveries opened the way to a radical new concept of a unified universe. The new conceptual paradigm goes under the name of grand unified theories or GUTs. Although Newtonian mechanics—a fragmented view of our universe—was perfectly adequate for over 200 years, it became an instant casualty of the "new age of physics." Whereas Newtonian physics separated and alienated man from the universe, the new physics established a perspective of quantum wholeness.

From time to time, a true genius emerges. Albert Einstein was such a man, the perfect innate consciousness. Einstein, like all true geniuses, was born with a capacity for cosmic connection. Unlike spiritual-seeking individuals, he did not have to strive for connection. But the questions that must be raised are, "Did he or did he not reveal something that had previously existed? Did he invent anything new? Have Einstein's fellow scientists, by exploring the structure of the universe and devising new technological tools and devices, radically altered the state of existence?" No. A genius, far from being the initiator of new concepts and inventions as commonly believed, is actually a channel for the cosmic unity.

At any given period in time, for some specific reason, when cosmic intelligence is ready for revelation, someone will be chosen for the job. A particular intelligence describing an aspect of our already existing universe must now become manifestly expressed. The questions about: "Who and why, and why now?" are inevitably linked with the concept of reincarnation. Who? Someone through whom this new intelligence will be accepted. Why? To provide another link toward ultimate enlightenment. Why now? Based on humankind's activity at that precise time or some radical metaphysical revelations, our giant cosmic computer system makes contact with the new components and the information is released.

This intellectual energy, in truth, is not born from the individual revealing the information but rather from the enormous input of the collective activity of the human race or metaphysical energy-intelligence input within the cosmos. In 1905, Rabbi Yehuda Ashlag, founder of the Research Centre of Kabbalah in Jerusalem, decoded the mystery of Rabbi Isaac Luria's theories on relativity and parallel universes. Not by coincidence, *only then* did science, in general, and Albert Einstein, in particular, begin *their* revelations of the general relativity theories.

However, at the time of Rabbi Ashlag, the teachings of Kabbalah were not widespread. Had Kabbalah, in fact, been as accepted worldwide as it is today, the idea of fragmentation or fission might have been easily recognized as negative and hazardous to the well-being of Earth's inhabitants. However, since the 1960s, millions of people have become familiar with the kabbalistic view of our universe. The study and scanning of the *Zohar* has pervaded and impregnated the cosmos with a "light of reality," like a lighted match in a dark room.

The universe, declares the *Zohar*, is a collective or wholistic concept. There is clearly a subtle link between the reality of the microscopic or metaphysical world and the familiar, physical (illusionary) macroscopic world. Therefore we cannot separate the quantum reality from the structure of the entire universe. The state of an individual revealment—whether it be fission or fusion—is meaningful only when it is regarded in the context of the whole. The microscopic (metaphysical) and macroscopic (the illusionary physical) universes are interconnected and interwoven as one fabric. They can never be separated.

This idea that a non-perceived, wholistic order exists in the universe by no means originated with modern physics. The Kabbalah is the *first* and *only* authority to present the doctrine of a cosmic order in which the affairs of humankind are reflected in the organization of the cosmos and then beamed back—as a satellite—to become manifested in strict physical terms and states of revelation. The revelation of first fission and then fusion are perfect examples as to how and why they became ultimately revealed in our world.

The negative activity of mankind invaded the cosmic structure and activated the universe of the Tree of Knowledge. The result, beaming back, was Einstein's revelation of atomic *fission.* The Tree of Knowledge universe contains the essential features of decay, chaos, and disorder. Thus, there is little wonder that the fission consciousness of humankind must, of necessity, bring upon itself the hazardous consequences of pollution, acid rain, and nuclear waste. These horrors represent the chaotic components of the Tree of Knowledge universe.

Negative energy-intelligence consciousness has been humankind's trademark since the Fall of Adam.[115] The conceptual problem, at the center of universal chaos, is the confusion between the processes and the origins. Instead of asking why illness, pollution, and misfortunes occur and trying to remove the conditions leading to them, researchers try to understand the ways by which the chaos operates so that they can then invade and interfere with it. The same is true of government and politicians who tend to be blind to the *origins* of disorder and conflict. They concentrate instead on the external processes, on the visible acts of crime and violence, rather than the concealed, metaphysical elements that bring about the evident chaos and violence.

It is vital for our well-being that we now change the situation. However, this revolutionary process is only possible if we are able, as a society, to shift to the new quantum, wholistic model. A wholistic approach must consist largely of a review of that network of patterns, be it economic, social or political, out of which conflict, strife, and fragmentation arise. The origin remains with human activity.

Let us now briefly examine the *Zohar*[116] upon which the idea of quantum, some 2000 years ago, was presented. It is now time to apply the kabbalistic techniques at all levels—individually, nationally, internationally, and globally.

> *"And if you shall say, 'What shall we eat in the seventh year...'"*[117] *Rabbi Yehuda opened the discussion with: "Trust in the Lord, and do good; dwell in the land, delight and pasture in emuna [from the word Amen]."*[118] *For when one clings to Supernal Emuna, so as to be connected with the Lightforce, no one in the entire world can cause harm unto him.*[119]

> *Come and behold: What is meant by "and do good"? A deed below, arouses an activity above; when you perform a good deed below, then that good of above, known by the code word Zaddik [righteousness] is aroused. Then you may "dwell in the land" with confidence, eat of its fruit and delight in it.*

> *However, if one does not arouse Cosmic Zaddik by positive actions, then Zaddik distances itself from the person, and then the land becomes as a burning furnace, and the fire burns to consume the world... Hence, if you shall ask, '"What shall we eat in the seventh year?" The answer is, "I will shed My blessing upon you in the sixth year," just as is written elsewhere "See that the Lord has giveth you on the sixth day the bread of two days."*[120]

There is no doubting the strong metaphysical elements that underlie much of the new perspectives concerning our world. Time warps open up rich possibilities in the belief that there is more to the world than meets the eye. Especially attractive is the strong wholistic flavor of alternative medicine.

The *Zohar* is quite clear in its determination that man and the cosmos are mutually supportive and inseparable aspects of the one all-embracing unified whole of reality. You cannot have one without the other. There is a unity to the universe that speaks out and says that without considering everything, our conclusions are nothing.

There is yet another lesson presented in the preceding *Zohar*. We have been educationally programmed, media-bombarded as follows: "When you have a headache, take something for fast, *temporary* relief. When you have financial problems, pray for the necessary funds to come your way." What we seem to overlook in our approach to life is that our requests are always presented in a fragmented manner. We are not conditioned to ask for *everything*. So we always settle for less. Does this mean that if we were to make a request that all our needs, the known and unknown, be met and fulfilled that the good Lord will grant our every wish? The answer from the *Zohar* is an emphatic yes.

The reason mankind has not made much progress in this direction is because we exist in a fragmented framework. Consequently, we are not accustomed to thinking or acting in a manner of quantum. This requires a *whole* new revision of global consciousness and should be considered as a task nothing short of a universal revolution without violence. The *Zohar* is a work with the potential to radically change our lives. The *Zoharic* teachings contain compelling explanations of why so many things seem to be going wrong in the world.

The energy-intelligence power contained within the words of the *Zohar* can make it happen. This assurance is given us by Moses, the master teacher himself. The awesome Lightforce issuing forth from the *Zohar* can turn our society around from a hard, fragmented, mechanistic people to a soft, wholistic-oriented community. Rabbi Shimon bar Yochai[121] maintained that after the ravages of time and disintegration, there will come a turning point. A powerful beacon of Light that has been concealed will return in the Age of Aquarius. There will be movement but it will not be brought about by violence or force. The old ways will be thrust aside and the new ways reintroduced.

We have very real and serious crises that must be dealt with and yet there are no solutions in sight. The basic thesis of this book is that our environmental disasters and the rising universal wave of crime and violence are all essentially a crisis of human negative activity, brought on by a crisis of fragmented perception. Like all other crises facing governments today, it derives from the fact that we are applying concepts of an outdated world perspective—the fragmented world view of Cartesian-Newtonian mechanics—to a reality that can no longer function in terms of these concepts.

The problem is further aggravated by the world becoming more globally interconnected, thrusting different cultures together, hoping and trusting everything will work out. Mistrust and suspicion, hastened by value differences, simply will not disappear because we are *told* that a new vision of reality, a change in our perceptions and values must take place. A paradigm shift cannot, and will not, take place until such time as the cosmos has been cleansed of all human created waste and pollution. The *Zohar* has made this perfectly clear.

So where do we begin? The various manifestations and implementations of this "new" paradigm are the subject of this book. The 60s and 70s have generated a whole new category of movements that still continue to the present time. Revolution, unfortunately at times violent, and rebellion, sometimes misguided, are wearing through the fabric of all peoples and nationalities.

So far, most of these causes that begin with a great deal of enthusiasm seem to run out of steam. The original founders finally withdraw when the results fail to meet their expectations. The underlying reason why these movements disappear is their failure to recognize the quantum cohesiveness between all peoples and the cosmos. A local or national movement will not succeed in achieving its objectives as long as these causes retain their fragmented, regional perceptions. We will, and must, learn that we are a wholistic society, where anything, everything, and everyone is affecting the individual person.

This statement does not imply that we must lose our identity or individuality in response to the quantum, interconnectedness factor. Each person, each nation, represents a different aspect of the all-embracing whole. But how can we provide a coherent conceptual framework that will help us recognize the communality of our objectives? Let's face it, for the present, it's every man for himself. Raising our consciousness to this new paradigm is a goal that few believe is attainable, at least in the foreseeable future.

The only item that all of us agree upon is that some change is necessary. The question as to *what* change is needed and more importantly, *how* to implement any new conceptual framework, is where the disagreements begin to appear. The truth of the matter is that most of us have come to the conclusion that there is really no hope for the future, so we might as well take full advantage of what is left of the present.

This book has raised some serious doubts about our future. It is by no means a critique of past performances of individuals, corporations or governments. Since time immemorial, these same conditions have existed, and they have been replaced only to bring about the same failures as the past. The *Zohar*, now with us for some 2000 years, can provide some of the missing ingredients that our previous societies have lacked. The essential factor that can, and will, change all societies within our universe is an additional infusion of positive energy-intelligence as clearly stated by the *Zohar*. With this positive infusion into our universe, humankind, the determinator of how our world behaves—pure or polluted— will have a better chance of creating positivity.

We are so inundated by the overwhelmingly negative attitudes that presently pervade and weave through the very texture of our society that it becomes difficult, even for those who have a natural desire to share and respect their fellow man. The cosmic atmosphere impels us to overreact in situations where, if the climate were more conducive, we would all most certainly behave in a positive manner befitting the higher species of the Lord's creation. Once again, fusion rather than fission.

This same idea has already been expressed by the famed kabbalist, Nahmanides, Rabbi Moses Ben Nachman, 1194–1270, also known by the acronym Ramban, in his commentary on the biblical verse: "And the Lord said unto Moses, 'and I shall harden the heart of Pharaoh.'"[122] Nahmanides questions the meaning and implication of the verse, inasmuch as it appears to be contrary to the doctrine of free-will and determination. If indeed, questions Nahmanides, Pharaoh decided not to remain a passive witness to a situation that he already regarded as dangerous and unpleasant, why did the Lord invade his space and framework? Why was Pharaoh not permitted to exercise free-will[123] and decide to free his Israelite slaves? Only a fool could remain oblivious to, and make mockery of, the plaguing events that were eating away at the Middle Kingdom Empire. If present trends were to continue, impending doom and eventual collapse were a certainty.

Important changes had taken Pharaoh by surprise, and it was difficult to control the manifestation of the plagues. In contrast, the possibility of what was to come, namely the severity of the plagues, increased. The likelihood of disaster was certainly convincing enough, even to an ego-filled emperor.

Consequently, Nahmanides, faced with these insurmountable objections, came to interpret the above verse in the *Book of Exodus* with the following exegesis. "Hardening of Pharaoh's heart" was simply a process of restoring cosmic influence to the time preceding the plagues. Pharaoh's present frame of reference was replaced by his original thought consciousness when plagues did not interfere nor invade his framework. He proceeded through a time-tunnel back to when he saw himself as ruler and mighty emperor of his Middle Kingdom Empire.

Now Pharaoh was faced with the decision of whether to free his slaves at the behest of Moses or demonstrate the awesome power of rulership. At this time, however, his experience with defeat played no active role. The suffering and hardship he and his entire kingdom had undergone played no decisive role in present and future decisions. This might be compared to a cassette tape where a particular section has been spliced and placed in animated suspension. In the human brain, the event is not lost to oblivion as the spliced section still remains in the "hard disc" of the brain and can be restored to conscious memory at any time.

The operations of human memory are extraordinary. Virtually everything one experiences in life is put into memory. These buried memories are astounding in their logic and meaningful strings of associations. The operations of the brain keep the data in orderly sequences and are arranged for usefulness and relevancy. And yet, despite all the available research, scientific knowledge cannot account for how different bits of memory become associated in a coherent manner or how the chaining of memories occur.

The elegance of our memory retrieval systems is so mysterious that we have only the faintest appreciation of their operations. Scientific research can describe many factors that influence memory and recall. However, science has little idea about how an intention to recall something can locate the appropriate memory, and at other times, the recall process draws a blank.

Yet the unconscious mind has its own reality and substance, however ethereal that reality might be. One of the curiosities in modern science is how something in the mind prevents immediate recall of bits of information that are on the tip of our tongue buried in the unconscious, until the required word or phrase struggles up from memory, and presto, the information appears. Something in the mind must know how and when to refresh our memories, while at other times we cannot remember at all.

Within the realm of human thought, scientific knowledge is most meager. The phenomenon of consciousness is perhaps the ultimate magical property of mind that defies understanding. It is a battered concept. The *Zohar* does, of course, know the necessary conditions for recall or relapse of memory. The code is "fusion" or "fission." This book does not deal with the entire realm of mind, but the *Zohar* [124] provides us with the clues about how the mind recruits the memory bank to perform its wishes, and how the mind can abstract the essence of an experience. Fusion or fission determines the extent and type of recall.

Because Pharaoh's power and consciousness underscored the element of fission or fragmentation, concentration by the

Lord over sequential events, causing them to become fragmented, was no problem. To lift a section or byte of information from Pharaoh's memory bank was not an unusual feat. Because Pharaoh had an awesome control over, and connection with, the Desire to Receive for Oneself Alone, the underlying cause of fission and fragmentation, the Lord "hardened Pharaoh's heart" by preventing any recall of the preceding events.

Pharaoh was now thrust into a situation where his previous negative energy-intelligence would not influence his present decision. There was a balanced cosmos—neither negative nor positive. Free-will for the Egyptian ruler was restored. The physical function of the human body could now freely express the mental activity of thinking, which in the case of Pharaoh was negative. Pharaoh was allowed the free expression of his desire without recognizing the results that his negativity would bring upon his house and his people.

If nascent creation were to stand any chance at all of surviving under the vital,[125] though terrible, burden of man's free-will, the universe and the world that would be mankind's had to be created with great care. The essential purpose of reading and scanning the *Zohar* is to restore the whole of the cosmos to its natural balanced state. Fusion is the fundamental thread weaving itself through the *Zohar*.

Only the *Zohar*,[126] with its construct in Aramaic, would be fit as the tool of the Lord's thought in the final days of Aquarius. We wait for society as a whole to behave morally and responsibly, and to repair the ecological damage we have created. Thus far, the steps we have taken to remedy our sick environment have failed miserably to materially affect any change in our well-being. We all overwhelmingly agree that the majority of today's emotional and medical problems have their origin in the many stresses of life. The awesome energy-intelligence power of the *Zohar* is the most direct method in providing effective solutions for solving problems. The *Zohar* can be a master healer. The problem facing us today is, "How can we learn more about its teachings so that we can heal ourselves and enhance our well-being?"

Scanning of the *Zohar*, at the very least, combines the fusion intelligence of its Aramaic writing with the computer-like mind. Many brain scientists compare the brain to a computer, while neglecting the obvious, namely that something outside the computer decides what to ask the computer to do. The conscious mind activates some quality related to this concept in the scanning process of the *Zohar*. Energized by the energy-intelligence of the *Zohar*, our mind-computers then tap the certainty paradigm of our universe. The results are astonishing.

In conclusion, however, I must repeat that the process and procedure of tapping the awesome power of the cosmos requires a consciousness of fusion attitude. Two of the most crucial questions we face in trying to understand how our mind and attitudes affect our body and overall well-being are, "Why are we not aware that our internal bodily functions are malfunctioning, long before we feel real pain or discomfort?" and "Why must it take 15 or 25 years for a doctor to tell us about a present, emerging condition that 'suddenly' became observable?"

The culprit is fission. Our lack of awareness is seen in every variation of human experience. If we can discover why we have poor perceptions of our internal functioning, then we will be more than halfway on the road to penetrating the awareness of our consciousness and preventing the breakdowns in our well-being. The teachings of the Kabbalah are to the individual what the sextant is to a navigator. To make it work requires a fusion attitude.

Chapter Five

STRESS

Chapter Five

STRESS

IT IS UNIVERSALLY AGREED THAT THE MAJORITY OF TODAY'S mental and physical problems have their origin in the stress of life. Yet nowhere is there to be found more confusion, misinformation and lack of direction concerning both the causes of stress disturbances and their treatment than in the therapeutic worlds of psychiatry and medicine.

While medical investigation of the stress syndrome has led to increased understanding of its central role in all aspects of illness, the unusually large incidence of heart disease in North America illustrates the disappointing results in dealing with stress-related diseases. Today, more than ever, there is an overwhelming sense of threat to life. Environmental deterioration, crime, and violence in the streets, and drug addiction, have all proven too much to handle. Law enforcement officers are feeling the stress of fighting a war they cannot seem to win. Our schools and neighborhoods have become open-air drug markets. For the first time, teenagers, our future generation, have topped adults in the percentages of serious crimes committed per capita.

Stress has become the target that plays a crucial role in just about everyone's dilemma. Ulcers are probably the classic example of a psychosomatic disease. Persons in stressful professions are likely to develop ulcers more frequently than others. This does not necessarily mean that the ulcer malady belongs to the private preserve of executives. However, where lifestyles run at a slower, more peaceful pace, the incidence of ulcers and other stress-related diseases is less prevalent.

By-products of the stress syndrome include the way stress plays havoc with sleeping patterns. While asleep, we may be restless or we may find difficulty falling asleep at all. Then, of course while awake, we feel irritated and usually exhausted. Stress is responsible for the breakdown of our natural biorhythm, which we so urgently depend on for our health and well-being.

There are even those medical researchers who are prepared to go so far as to consider every illness somehow originating from stress-related problems. How do health practitioners and medical researchers approach the definition or the nature of stress? Presently, there is little in the way of definition describing the essence, characteristics or tendencies that provoke illness or the breakdown of our well-being. However, there are endless examples of what kind of things cause stress. But in the final analysis, they cannot explain or describe what stress really is.

When people express that they are suffering under the pressure of stress, what they are *really* talking about is a condition or situation in their lives that has placed them under a great deal of tension. Stress then is *tension*. Financial difficulties, family problems, job insecurity, and a host of other things that cause stress are defined by researchers as underlying reasons for stress but they are not a definition of what stress is.

This list is a description of the situations that trigger stress within us. Trying to define stress is obviously quite different from defining something of our physical (illusionary) realm. For things in the physical realm, we believe we can provide descriptions of their specific properties or characteristics through test tube experimentation.

Stress, however, remains very much in the realm of metaphysics. If worry is simply a problem-solving activity, then why does worry lead to emotional anxiety? Werner Heisenberg, the world acclaimed German physicist and Nobel Laureate, answered this question with his uncertainty principle. Simply stated, the premise of this principle is that we can never trust our senses completely, and probably what we see is a questionable reality. The fact that we have no certainty in this world is what creates worry and anxiety for most of us.

Turning to our dictionary for a definition of worry raises more questions than satisfying answers. The standard dictionary describes worry as follows: "To be uneasy in the mind; to feel anxiety about something." This is usually the way we think about worry. Does this imply that if we experience a situation of "uneasiness," the result will be stress? If in our business there is doubt as to which direction is better, is this to be considered an "uneasy" situation? Flying in an airplane, for example, is a thrill for some and a nightmare for others. Some people worry when crossing the street. Others refuse to leave their homes for fear of what might happen once outside the security of their human prison-fortress. Then again, the fear of burglars entering our homes is also reason for worry.

What distinguishes the worriers from the non-worriers? Why do some people worry about things that others pay no attention to? Each of us is concerned with what the future holds in store. It is this uncertainty that pervades our minds and hence can cause anxiety and the like. But then again, what seems to be different in the 20th century that did not exist before? The uncertainty of yesteryear is exactly the same as today. No one escapes its grip.

Standard psychiatric practice interferes with the healing process by suppressing the symptoms. True therapy should consist in facilitating the body's natural healing process by providing an emotional, metaphysical supportive atmosphere for the patient. Rather than suppress the process that consists of symptoms, the symptoms should be allowed to intensify to permit a proper evaluation and prognosis.

The extension of the biomedical approach in physical medicine to the treatment of mental illness has been very unsuccessful. A tremendous amount of time and effort has been wasted trying to arrive at a basic diagnostic system of mental disorders without the realization that the search for this kind of physically-based diagnosis will ultimately prove futile for most mental disorders.

The Cartesian view of medical researchers often prevents psychiatrists from observing the beneficial aspects and potential value of so-called illness. Disease has come to be considered an enemy that must be dealt with and destroyed. Such a narrow point of view fails to utilize the subtle spiritual and psychological aspects of mental disorders, and consequently prevents researchers from arriving at beneficial, therapeutic methods for healing the physically and mentally ill.

It is, therefore, intriguing and most ironic that physicians themselves are the ones who suffer most from the Cartesian view of health. They completely disregard stressful circumstances in their own personal lives. Physicians live from 10 to 15 years less than the average population. They have not only high rates of physical illness but also remain high on the list for alcoholism, suicide, and other social pathologies.

We know that many of the professional theories about stress are illogical and often confusing. We live in a society completely dominated by the stress syndrome. Stress-related illnesses that did not exist prior to 100 years ago continue to increase at an alarming pace without showing any signs of abatement. This, in my opinion, is the direct result of physicians evading the essential issues: "What is stress and why does it affect some and not others?" The failure to adopt the kabbalistic approach to logic—asking *why*—is the fundamental cause for a society full of stress. We continue to use outmoded logic and procedures. We prefer the instant, temporary solution as opposed to researching the origin.

From a kabbalistic point of view, understanding the nature of stress and how to deal with stressful disturbances, does not begin with, and cannot ever *originate*, within the body's physiological mechanisms. Most clinical stress researchers claim that both the physical and emotional reactions that produce that so-called "thing" called stress are produced by an arousal of the body's mechanisms. They believe that when people are stressed, the body responds by activating what medical professionals consider to be defense mechanisms, such as an increase in the heart rate, blood pressure, and other body changes that rush to help the body defend itself.

The confusion increases as we explore the phenomenon of stress. Is stress, according to medical research, something that originates outside of ourselves, a result of some unpleasant condition? When we go to the movies and see a horror film, our heart begins to beat faster, our breathing then becomes irregular, and our blood pressure climbs. Is the movie the stress factor, and is distress the condition our physical bodies undergo following exposure to this horror movie? Is stress something internal that stimulates nerves that cause a chain reaction?

While medical professionals can identify the biochemical changes that occur during some external stress reactions, they still cannot explain how stress can activate the nervous system. Nor do they explain why our nerves act upon some parts of our body and not others. And furthermore, why are some affected by the horror film and others are not?

The physicists define stress as a "force" exerted on the body that tends to strain and sometimes even warp it. In other words, stress is something of an external event that wreaks havoc from without. This seems to be in disagreement with medical researchers who cling to the opinion that stress is our reaction to something that occurs within. This apparent contradiction, as we shall learn later on, comes from the

Cartesian split in medicine where there emerged two camps that have little communication with each other.

Physicians are concerned with the treatment of the body. Psychiatrists and psychologists are concerned with the healing of the mind. The gap between them results in a severe handicap in understanding the role of stress and emotional states in the development of physical and mental illness. The connection between mental states and cancer, for example, has been known for centuries. The kabbalist and the *Talmudist* report this link, which originated within our Cosmic Code, the Bible, long before the medical profession evolved as it is today.[127]

The reason for the lack of communication between the physical health professionals and psychiatrists is essentially the lack of confidence physicians have in treating patients on a thought or metaphysical level. They see the biological model as the basis of life. Mental activity is a secondary phenomenon, if considered at all.

There is, therefore, little wonder that the stress factor in medicine remains an often-confused, misunderstood phenomenon rejected outright by most physicians. Avoidance of the spiritual nature of the mind has characterized our modern technological society, and is reflected in the fact that treatment is intended to control symptoms and not to cure the illness itself. Consequently stress, a metaphysical component of man, is neglected by today's Cartesian science of medicine.

We have now established the distinctive nonphysical quality of stress. The so-called psychological stress of unhappy

TO THE POWER OF ONE

relationships, treated by many professionals as stress caused by unsuccessful interactions, is from a kabbalistic world view, still to be considered symptomatic in origin. Stress originating from external, physical causes, such as conflicts with the boss, mother-in-law, or colleagues are *not* by and within themselves inherently stressful. A mother-in-law is not an inherently stressful relative.

We target people—mothers, fathers, business associates—as the source of psychosomatic sickness. But from a kabbalistic perspective, these targets cannot be considered the origin of stress. According to the Kabbalah, it is an extension of the biomedical approach to physical illness to consider stress to be a situation where the mind misinterprets an outside incident as unfavorable or dangerous. Intelligence, mind, and brain are also components of an integrated physical, material body system. As such, a diagnosis of only the physical body will ultimately prove futile in determining the cause of stress.

These factors are not the critical events in stress and stress reactions. The intellectual systems of humankind interacting with personal performance and emotional activity, along with the myriad other mysterious functions of the mind, when creating discomfort, anxiety or worry as an expression of stress reactions, cannot be included in the kabbalistic definition of stress. This commonly understood definition is not a valid one.

The answer to the question of how and why we react to stress in the particular way we do must, of necessity, be found outside the physical, illusionary, material realm. When you, my readers, become acquainted with the now-famous uncertainty principle of Werner Heisenberg, the myth surrounding stress will gradually disappear. Nature, he claimed, always draws a veil across her face whenever a scientist makes an attempt to obtain precise information.

Heisenberg produced his own language using sets of conceptual tools to describe reality and its parallel counterpart, the physical illusionary reality. He believed that all physical reality, which includes subatomic particles, change at the whim of the human observer. And since the quantum reality implies that we must take virtually the whole universe into account when seeking the true cause of any physical event, we might as well say farewell to physics, namely to physics as it has been practiced for the past several hundred years.

To further confuse the issue, the famous 20th century Danish physicist Niels Bohr said, "If you don't see it, it isn't there." But surely the world out there exists whether or not we are looking at it. The chair upon which I am sitting may have disappeared sub-atomically but I am not likely to fall to the floor. The answer lies in the consciousness of the mind. The mind element appears to be essential to our observation of the real world. In other words, because we have decided that we are seated on a chair, the physical chair responds to our consciousness that a chair exists.

There is no question that this sort of thinking about the physical, illusionary world is new to the western mind, and more so, to the layman. In fact, all of this can be thoroughly confusing for the layman. However, we must refer to it to illustrate how unfamiliar we are with this physical world. After all, if stress comes from my mother-in-law, then the uncertainty principle and Neils Bohr suggest that I, in my mind, should consider that my mother-in-law has now departed for the North Pole, without access to a telephone.

Touching upon a more serious note, we may ask of ourselves, if the physicists—Neils Bohr and his colleagues—are indeed correct in their conclusions concerning reality, then why can a person with cancer not send signals along nerves to enhance defenses and make the body fight more aggressively against disease? Why can the patient not deal with illness the same way the physicist claims we can relate to a chair? "If you don't see it, it isn't there."

The point is that there is a great lack of communication between the scientists who investigate the essential building blocks of nature and the medical researchers. The information flowing from the physics establishment must relate to the medical field as well as to the layman. Until recently, scientists have not paid much attention to links between psychological factors and the body's immune defenses against serious illness. They have often discouraged the notion of such interconnectedness, despite increasing research showing that the mind plays a vital role in health.

If we take the physicist's claim, "if you don't see it, it isn't there," one step further, the subject of stress becomes even more confusing. How do reactions to stress begin? Why should perceptions of social relationships ever reach an unfavorable stage? The uncertainty principle already takes us directly to the consequences of unpredictability and thus of questionability. Suddenly, it can be seen that events that occur are not necessarily perceived with accuracy. Therefore, perceptions, favorable or unfavorable, are really a figment of our imagination.

If stress, as many maintain, is an imbalance of the organism in response to environmental influences, then its consequences should be considered uncertain. Why do these consequences become harmful? Furthermore, despite numerous studies that seem to indicate that prolonged stress suppresses the body's immune system, this link between stress and illness must, of necessity, be considered illusionary. Full recognition of the uncertainty principle will bring about a major shift in medical research. No longer will the preoccupation lie with symptoms and probabilities, but rather with a careful study of the origin of stress as encouraged by the kabbalistic perspective of reality.

Such a shift is urgently needed. The degenerative diseases that are distinctive of our time, and constitute the major problem of disease and death, originate and arise out of excessive stress. At this point, we lack information as to the origin of stress. To arrive at a reasonable answer or to engage in problem solving from bits of valid information or worse, misinformation, does not begin with a flight from reality. The need to become informed has become a part of therapy. Self-help together with awareness techniques can prove to be as useful as or even better than drugs. Stress is no more dependent upon the existence of the brain than a musician is dependent upon the existence of his violin, although both instruments are necessary for expression in the physical world. When we fully grasp this viewpoint we can then begin to approach the study of stress, with meditation as its therapeutic instrument. Soul-consciousness existed before birth. This is the first fundamental fact of reincarnation.

Before we begin our investigation of the intimate interplay between the physical and the Lightforce process, the term "psychosomatic" needs some clarification. In conventional medicine it was used to refer to an illness or disorder for which a clearly diagnosed physical basis could not be established. Consequently, only symptoms could be treated and most of the time the disorder returned. Despite the extensive literature on the role that the mind plays in influencing the development of illness, very little time has been allocated to exploring the origins of these influences or the methods of altering them.

The clue to any such attempt lies in the fact that these forces and processes not only play a significant part in getting sick but they also should and must function within the healing process. The first step in this kind of self-healing, an integral part of the study of Kabbalah, must include the patient's recognition that he or she must consciously participate in understanding of the *origin* and development of the illness. Without this awareness and consciousness, the patient will fail to make a connection with the healing process.

What we do not "know" is that with which we have not yet come in contact. "Adam knew Eve and she conceived and bore Cain."[128] An obvious difficulty with this verse of the Biblical Cosmic Code is, "How can the mere act of knowing create a pregnancy?" The *Zohar* explains this simply as the difference between information and knowledge. Knowledge is the connection. Knowledge is energy-intelligence. "Obviously there was an act of physical intercourse between Adam and Eve," states the *Zohar*. "But why does the Bible make use of the word knowing to indicate sexual intercourse when other suitable words in Hebrew are available?" "The verse in Genesis," continues the *Zohar*, "indicates that when there is knowledge, then a direct contact has been established."[129]

Consequently, when people are ignorant as to the origin of their illness, they must, and usually do, abdicate all freedom of control over their lives. Conditioned as we are by the Cartesian framework, we may even refuse to consider the possibility that we could have participated in our own illness. We then freely pass control over our lives to the physician or psychiatrist, in whose hands we blindly place ourselves. One of the primary difficulties in kabbalistic techniques lies in the concept of patient responsibility.

Knowledge, as I previously mentioned, is essential for the healing techniques of Kabbalah. It is for this very reason that I have written so extensively about the basic mechanics. Kabbalistic techniques make demands upon its practitioners. But then again, from a kabbalistic perspective of our world, we were never destined to behave as anything more than a complex mechanistic living organism.

In physics, the mechanistic paradigm had to be abandoned at the level of the unobservable. Transcending the biomedical model will, and must, amount to a major revolution in medical science. Because of the influence of Aquarian Age, a new vision of reality, a fundamental change in our perceptions and values, is likely to dominate over the entire planet. Man is once again being called upon to take hold of the reins of a galloping, runaway universe. The violent and reckless behavior of society demands that rule over our destiny return to the people alongside proper sensible government.

In no way does this statement imply that government, under the given circumstances, is not trying its best. Faced with ever increasing dangers to our ecosystem—air pollution, toxic food, and water—experts in various fields of government can no longer deal effectively with the urgent problems that have arisen in the past decade. The "brain trusts" and "think tanks" of yesterday admit they are unable to solve the world's most urgent problems. In their own confusion and defense, they usually cite the preponderance of new situations, new circumstances, along with irrelevant changes in outdated conceptual models.

Let us, therefore, now begin our investigation of this phenomenon known as stress from a kabbalistic world view. Unlike previous cultural fluctuations and declines, the crisis we are facing today is no ordinary crisis. The great transition phase that mankind is now experiencing will be more dramatic than any of the preceding cycles in human history. These dynamic transformations do not take place very often.

According to the *Zohar*, there have been few transformations that have taken place in the recorded history of man, and among them are the rise of Judaism with the Revelation on Mount Sinai[130] and the Israelites exodus from Egypt.[131] The transformation we are running up against now may well be more dramatic than any of the preceding ones. The questions confronting us are, "What are these changes about?" and "Why *now*?" Because the transition is more extensive and involves the entire universe—terrestrial and celestial—it is not the basis or cause for our present upheaval. As I mentioned on many occasions, events expressed on a physical level can never be deemed a cause. All physically manifested expressions are merely extensions of a thought, metaphysical energy-intelligence consciousness. Thought or the unknowable remains the primary cause for events occurring on our globe. Facts or incidents of a physical, observable nature are secondary occurrences that result from primary causes.[132]

A clue to the answers to the questions raised is provided by the *Zohar*[133] in a discourse on the Age of Aquarius:

> *Rabbi Shimon raised his hands, wept and said, "Woe to he who meets with that period; praiseworthy is the portion of him who encounters and has the divine capacity to be cast with that time." Concerning that time it is proclaimed, "And I will refine them as silver is refined, and will try them as gold is tried."*[134]

The *Zohar*[135] also states:

> *... in the days of the Messiah, there will no longer be the necessity for one to request of his neighbor "teach me wisdom," as it is written: "One day, they will no longer teach every man his neighbor and*

> *every man his brother, saying 'Know the Lord,' for*
> *they shall all know Me, from the youngest to the*
> *oldest of them.*"[136]

The *Zohar* here expresses the idea that the Messianic era will usher in a period of unprecedented enlightenment. Messianism, representing the essence of hope and optimism, is bound up and dependent upon true knowledge, the sublime wisdom of the Kabbalah. With worldwide dissemination of the *Zohar* expanding at an ever-increasing rate, the fulfillment of Jeremiah's prophecy, "they shall all *know*," is closer to reality than ever before in history. Knowledge is the connection to the whole of the cosmos. Why, we might ask, is knowledge so essential in this Age of Aquarius?

Knowledge, originating within the software of the *Zohar*, permits mankind to access into quantum knowledge of the universe. The possibility of raising our level of consciousness is assured by installing this universal software onto our own particular *mind* computer.

Knowledge, from a kabbalistic world view, is energy-intelligence more powerful than any energy system employed for energy purposes. Knowledge permits mankind to relate and make the proper circuitry with the awesome power and energy of the cosmos. Knowledge of the *Alef Bet, Zohar* and other highly charged kabbalistic materials prevents a "burning-out" of our physical and mental systems. The energy-intelligence established by kabbalistic circuitry is the very stuff that can fuse the energy of the cosmos—more powerful than the energy of the sun—with human beings.

The question that must be raised at this time is, "Why did the Prophet Jeremiah consider *knowledge* essential during the Age of Aquarius as opposed to other significant periods in recorded history?" The Age of Aquarius (the time of the days of Messiah) will usher in an unprecedented infusion of Light energy-intelligence as indicated by the previously cited *Zohar*.[137] This might be compared to the entire universe living adjacent to high tension wires where the energy fields are intense. Without the proper human circuitry channels permitting an even, smooth flow of this awesome Messianic Light energy-intelligence, mankind will remain in a state of electrical shock all day. This is similar to having our finger affixed to a live socket all day without any possibility of withdrawing from it. This condition will be constant without any relief, not even temporary relief.

The *Zohar*, *Vayikra* 59:385-388, explains,

> *Come and see, as long as Moses was alive, he used to check Israel from sinning. And because Moses was among them, there shall not be a generation like that one till the days of Messiah. How much more those who stand before Rabbi Shimon himself, who is above all.*

> *...Alas for the world when Rabbi Shimon will depart from it, and the fountains of wisdom shall be sealed from the world, and men shall seek words of wisdom, but there will be none to impart it, and the Torah will be interpreted erroneously because there will be none who is acquainted with wisdom. ...*

> *Said Rabbi Yehuda: The Lord will one day reveal*
> *the hidden mysteries of the Bible at the time of*
> *Messiah because "the earth shall be full of the*
> *knowledge of Hashem, as the waters cover the*
> *sea."*[138]

Towards the arrival of the Age of Aquarius, the *Zohar* holds
out more hope than a science that must rely largely on
randomness and probability. The *Zohar*, also known as the
Book of Splendor, provides a direct link and contact with
the universal energy-intelligence that we discussed
previously. The science of the Kabbalah does indeed answer
many of the enigmatic aspects of nature, yet it still remains
elegantly simple.

What lies in store for us is an overload of Aquarius energy-
intelligence, more intense than any other that we now
experience. The reading or scanning of the *Zohar* will create
the adjustable, letter-right personal receptacle that provides us
with the ability to receive this awesome power of energy
without being burnt-out in the process.

The question that might now be going through your mind is,
"How in Heaven's name can *I* connect and make use of the
Zohar. I have not the faintest idea how to read the material.
And even if *I* could read the Hebrew letters, *I* don't have the
slightest idea what it means?"

For those of us who consider mankind beneath the intelligence
of a computer, there does exist a problem.

However, if there are some of you who consider yourselves at least equal to the intelligence level of a computer, then I make a suggestion: Take a trip down to your supermarket and observe the check-out counter. There you will notice the clerk passing a purchased item over a scanner with a funny looking squared configuration known as a product bar code that usually appears on the back of your product facing the scanner. The cashier will tell you that the scanner immediately relays this information to a computer which, in turn, transmits the purchase price back to the cash register, almost with the speed of light.

This is precisely the interrelationship that exists between our eyes and our mind computers. The scanning of the *Zohar* immediately establishes the information in our mental software. It can be compared to transferring information from one floppy disk to another. We now possess the software program by which to connect with the cosmic hard disk of reality where certainty, yesterday, today, and tomorrow are all one. Decay and chaos do not exist within the reality realm.

To enhance our understanding of the Aquarian Age, we must ask, "What happened to the much-touted Age of Aquarius, with its promise of harmony and understanding, sympathy and love abounding?" Never before have we hovered so close to the brink of nuclear disaster.

According to kabbalistic wisdom, the physical world is just a blip on the endless screen of reality, a temporary static disruption, a minor disturbance of the endless peace. The space-time physical reality that mankind experiences is merely a pattern of interference that exists only for the flash of an instant where we have lived as physical entities. This physical reality will be here only until that time, at the end of our process of correction, when the universe fine-tunes itself out of existence.

What proof can the kabbalist offer to substantiate such seemingly outrageous claims in light of the fact that we seem no nearer to resolving our problems than did our primal ancestors? Indeed, if anything, the situation seems to have gotten worse. How can the kabbalist's faith and optimism remain undisturbed with the specters of war, terrorism, and nuclear proliferation looming like dark clouds on the horizon of human consciousness?

The kabbalist sees the struggle of science to achieve more with less as a reflection of man's striving to shed his garments of darkness and step once again into the Light. Thus, these developments, when seen from a kabbalistic perspective, reveal an inborn tendency in man to strip himself of the stifling raiment of physicality and to embrace the Infinite.

The Light never rests. It is forever compelling us toward the culmination of the cosmic process, the Final Correction. In this Age of Aquarius, It incessantly urges us on toward that heightened and intense state of consciousness. With each passing day, the energy force of the Light intensifies in ever increasing stages to provide us with an opportunity to remove negative energy-intelligence and end for all time the need for Bread of Shame.[139]

The greater the Light's revealment, the greater the pressure on us to reveal it. There is an old kabbalistic saying that, "As much as the new born calf needs to suckle, the cow feels a greater need to share its milk." What the kabbalist sees today is an increase in the pressure, a hastening of the connective process that augurs the beginning of the end of a long, arduous process of spiritual adjustment and rectification, and for many individuals, the dawning of a New Age.

Consequently, stress originates with fulfillment, with the Light, Whose desire it is to enhance and fulfill the desires of the vessel, of mankind. In this Age of Aquarius, the Light no longer curtails its Desire to Share, nor does It consider the present dimension of our capacity to receive fulfillment. No. The Light now directs its fulfillment towards the full potential of each individual's energy-intelligence of desire.

However, if one is not instructed in the doctrine of Restriction[140] then a severe burnout must occur. Correspondingly, the individual whose practice of Restriction has become entrenched within his or her mental computer will reap the maximum benefits of the Aquarian Age.

Thus, the *Zohar* declares the paradoxical circumstances that shall surround mankind during the Age of Aquarius. "Woe unto those who shall be present in that time," refers to those incapable of containing the enormous flow of Light, whereas spiritually-oriented people will reap the benefit of the dawning of this New Age. The *Zohar* assigns the beneficial portion of its declaration "praiseworthy are those who shall be present in that time," to those people who have taken control of their Desire to Receive for Oneself Alone.

What seems to emerge from the foregoing is the startling kabbalistic revelation that *stress is the Light* struggling to fulfill mankind's insatiable desires. Unlike the common notion that stress is behind the numerous diseases that have sprung up in the past century, the kabbalistic view is quite the contrary. The Lightforce is the vital element necessary for enhancing our physical and mental well-being.

Medical science and man want to relieve the effects of stress. This is indicated by the widespread use of tranquilizers, narcotics, cigarettes, and alcohol. As a result, freeing the body from stress has become a frantic quest for even more fulfillment and pleasure. Material comfort and achievement do provide some degree of satisfaction but the state of fulfillment, or the lack of it, is what determines the overall quality of one's life experiences. If stress can fill a person's everyday activities with anxiety and dissatisfaction, then the commonly held assumption is that the relief of stress will result in the physiological opposite of stress, namely, a tranquil and dynamically functioning nervous system. However this has not been the case.

Excessive stress limits can even curtail productive decision making. Freeing the body from stress is mistakenly taken as a process that should reveal reserves of energy and intelligence. This temporary relief of stress, whether by the use of drugs or medication, can momentarily improve performance and effective, constructive thinking. We have been programmed from birth to the value of fast and temporary relief. However, with much chagrin and disappointment, we fall prey to an avalanche of unsuspected and sometimes overwhelming problems that we are not prepared to deal with.

Instability is the hallmark of a person in stress. The fast moving pace of our society demands that people *quickly* relieve their internal stress and begin to use their full faculties to resolve the crises created by increasing cultural and societal fragmentation. We must become exceedingly flexible. We must adjust ourselves to meet a wide variety of rapidly changing demands without incurring excessive stress. We are constantly on the run and preoccupied with moving into, and fleeing from, stressful situations.

It, therefore, should come as no surprise that today's medical statistics indicate that we are a society journeying from one stressful situation to an endless array of others. Perhaps we have embarked on the wrong path, one that sustains sickness in our lives. The possibility of our achieving a permanent solution seems as remote as our achieving improved health, emotional stability, and performance.

This book describes a method by which we can achieve this permanent solution by encouraging, yes, encouraging, accumulated stress. After all, acquisition of increased stress is one of the indicators that a greater degree of Light or fulfillment is destined for me. Shall I turn it away? Am I to refuse this wonderful opportunity for the enhancement and improvement of both my physical and mental well-being?

This novel idea seems so totally contrary to our inborn concepts concerning stress. And yet, let us consider a similar situation where stress and body come in conflict with each other. Let us assume we have decided to open a business that manufactures a product requiring heavy machinery. We must consider the stress load requirement of the floor holding the machines. Assuming that the machines exceed the limits provided by the floor structure, will our decision to continue the business mean the removal of the present machinery and its replacement by smaller, lighter equipment?

Only a fool comes to this kind of conclusion. The alternative solution is to shore up the floor to permit the existing equipment, or even heavier equipment, to serve the current and future manufacturing demands of the business. The machinery in this case is to be compared to the Lightforce, while the floor is to be compared to our Desire to Receive for Oneself Alone or body energy-intelligence. Restriction or resistance to the oncoming fulfillment by the Lightforce is the prerequisite for the expansion of our vessels.

The degree to which the Lightforce assumes a beneficial or stressful characteristic depends on our ability to direct and hold onto the reins of the Lightforce. And to the extent that all of us successfully handle or deal with its sharing, beneficent energy-intelligence, the rewards are translated into mental and physical well-being.

To explore methods by which the stress factor, meaning the Lightforce, can be reduced or even removed results in a never-ending cycle of recurring stress. The Lightforce is not impressed or influenced by our desire to become withdrawn from its energy-intelligence. Its fundamental desire is to share, and nothing we do can prevent the Lightforce from achieving Its objective.

It is essential that we understand the nature of stress. For most people, the Lightforce might appear as an enemy placing unwanted energy-intelligence upon them. The paradox is that fulfillment and satisfaction are the essence of the Lightforce. So while we may prefer to rid ourselves of the burden to shore up or expand our vessel by the doctrine of Restriction, we must face the inevitable consequence of a reduced Lightforce and subsequently a diminished fulfillment. By being in a state of diminished fulfillment, we can expect to be in a constant state of frustration because our natural Desire to Receive is being ignored.

So here we are caught between the devil and the deep blue sea. If we have not properly expanded our vessel's dimensions of receiving, then the Lightforce puts too much pressure on us, per our square inch. On the other hand, to reject the Light's beneficence fails to satisfy our natural Desire to Receive and become fulfilled with "everything."

The use of tranquilizers or other stress-relieving drugs is only a temporary expedient. In effect, what is really happening is that the body energy-intelligence is losing or being removed of its consciousness. When someone takes several alcoholic drinks, he or she forsakes and abandons the instinctive limited scope of body-intelligence, the Desire to Receive for Oneself Alone.[141] The soul then soars to freedom from bondage inflicted by the chains of body-intelligence. The soul, whose energy-intelligence essence is a Desire to Receive for the Sake of Sharing, has once again returned to and embraces the Lightforce.

The relief, *light-headedness* that is at once experienced, comes as a direct result of the soul embracing the Lightforce without the usual obstructions and limitations thrust upon it by body energy-intelligence. Body energy-intelligence has been numbed temporarily. It has succumbed to the onslaught of drugs, alcohol or other medications. The result is a kind of euphoria induced by these artificial properties at war against the enemy, our body.

The problem, however, is the return trip. The effects of artificial methods remain for a limited time only. The body energy-intelligence, waging a constant battle to fulfill its objective,[142] regains its consciousness. It then resumes the battle with the soul-consciousness for control over our activities. Our soul-consciousness was provided with a free ride to *rendezvous* with the Lightforce. Our souls have tasted the bliss of eternal Light. However, our body intelligence— the force of limitation caused by its characteristic Desire to Receive for Oneself Alone—*did not actively participate* in this joyride.

The infinite Lightforce has no relationship or communicative wavelengths with body-intelligence. The two of them are diametrically opposed to each other. They serve opposing purposes. The Lightforce, an extension of the Lord, knows only the attribute of sharing. The soul's essential consciousness is one of Desire to Receive for the Sake of Sharing. The body's function, however, is to instill the consciousness of Desire to Receive for Oneself Alone.[143]

Consequently, upon the soul's return to its material, corporeal place of lodging, the soul-consciousness, again, experiences the limitation of being unfulfilled invoked upon it by body-consciousness. Now, however, the dichotomy between soul and body has been widened by the soul's recent unlimited contact with the Lightforce. The body-consciousness continues on its path of divisiveness, fragmentation, and lack of fulfillment due to its essential characteristic Desire to Receive for Oneself Alone.

The Desire to Receive for Oneself Alone opposes the doctrine of quantum relationships. Its thought-intelligence, by its very nature, thinks about and pursues those activities that benefit itself alone. There is never any consideration beyond the scope and framework of self. Consequently, by its limited perspective, body-consciousness cannot, nor does it have, the desire to expand the dimensions of its vessel. This is indeed a paradox.

However, so go the laws and principles of our universe. By rejecting the Desire to Receive for Oneself Alone, just as with the filament of a bulb, the reverse objective is achieved. The brightness of a light bulb is determined by the size of the filament, not by the current that runs through the wiring system. The current is the same no matter what appliance is plugged into it, whether it be an air conditioner, with great demands, or a five watt bulb, with small demands.

Restriction, a concept to which body-consciousness does not subscribe, is a necessary response to the Lightforce. The very nature of body-consciousness is one of illusion and finitude. Its scope is limited and is in direct conflict with the soul's Desire to Share. So after the soul has been given *free* access to the Lightforce, upon its return it finds the body more repulsive than ever before. Caught up in this worldly struggle for survival, the soul wishes to return again for an ever refreshing stimulation by the Lightforce.

The Lightforce wishes to embrace the soul but is prevented only by the doctrine of "no coercion in spirituality." The soul, whose intent is to become unified with the all-embracing unified Lightforce, must first meet the requirement of transforming the body's consciousness from a Desire to Receive for Oneself Alone to one of Sharing. Thus, the internal struggle within us becomes more intensified. The alternatives and solutions are few.

We have few options open to relieve the crisis of stress. We must create a transformation of the body-consciousness from one of Self Alone to that of Sharing. In this process, consciousness of the body undergoes an expansion of awareness and receptivity. And as we have seen, the Lightforce now enjoys a greater affinity with the consciousness of the body, and is permitted by cosmic law to increase Its share of beneficence to the body.

By increasing and shoring up the ability of body-consciousness to push back the Light (Restriction), the intensity of the Lightforce becomes vastly increased, further satisfying the desire and need of the body. The soul- and body-consciousness experience a greater affinity with each other, thus permitting the Lightforce to become an integral force within the psyche of man. The stress factor has been alleviated, if not entirely removed, by virtue of the Lightforce no longer being considered an outsider.

Although the Lightforce makes every attempt to forcibly have Its presence known and experienced within the consciousness of mankind, It is no longer considered an "overload factor." When our vessel is reinforced and reprogrammed, the psychological and physiological *opposite* of stress may be expected. The inevitable result is an improvement in the overall quality of our emotional and physical well-being.

If, on the other hand, we fail to improve the energy-intelligence of the body to one of Sharing, then we must go through the trauma and frantic quest for ever more fulfillment. So long as the body, the vessel, is *incapable* of embracing the Lightforce, a burnout or overload experience becomes a daily recurrence. The use of larger and more frequent doses of artificial repressors to relieve the effects of stress serves no other purpose than to increase the numbing effect on our body-consciousness.

The Light will not fade away. We must make room for Light's entrance into the consciousness of mankind. Connection with the Lightforce in the Age of Aquarius is no longer reserved for those who merit Its influence. All of Earth's inhabitants now fall under Its demand for revealment. The only choice we have is to increase our vessel to a maximum capacity, and embrace all of the Lightforce destined for us.

To reduce the ability of our Desire to Receive by artificial inducements does not influence the behavior of the Lightforce. The Lightforce will still make a continuous effort to fulfill "the Desires to Receive of mankind," and to improve mankind's physical and mental well-being. This is what the Lightforce is all about. Without Its beneficence, mankind must endure pain, suffering, uncertainty, and lack of fulfillment.

To numb the Desire to Receive by drugs or alcohol is to deprive people of their given birthright to enjoy the all-embracing whole of the Lightforce. Adopting any form of meditation only for the purpose of quieting a stressful body will have the same negative results as the use of artificial repressors. Both methods fail to adequately deal with the Lightforce.

The study of Kabbalah, on the contrary, describes a method of achieving fulfillment by *increasing* accumulated stress by the Lightforce. This also provides the body-consciousness with an expanded measure of integration with the Lightforce while achieving stability and fulfillment for the nervous system and each integral structure of our body.

The study of Kabbalah enables us to gain a deep state of rest, which repairs the damage incurred by previous excess stress because the action of the Lightforce vaporizes body obstructions. Those black spots that appear on X-rays are merely indications of the impasse created by the body-consciousness in hampering the steady flow of the Lightforce.

Increasing the level of the Lightforce promotes improved health and advances the emotional stability and performance of those who practice it.

Our failure to meet the demands of the accumulated tensions of daily living, brought about by the crisis of modern life, lies in our inability to secure the Lightforce within our consciousness. The widespread use of addicting and harmful medications for relief of excessive stress and related illness must be considered as a complication of, and not a solution to, the problem of stress. We must develop the full capability of harnessing the Lightforce to achieve and sustain fulfillment of life in the midst of these trying times. Without a doubt, the study of Kabbalah and the use of Kabbalistic Meditation, when practiced properly, can enhance our day to day lives in the most unimaginable ways.

The Ten Sefirot

The Upper Triangle

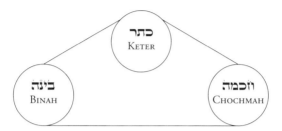

Seven Lower Sefirot
Seven forms and levels of cosmic
intelligence—the manifestation of the Lightforce.

These intelligence-coded messages (metaphysical DNA) account
for our grand solar system and for the Earth's cosmic division.
The Seven Sefirot are encased in the heart and soul of the planets.
The shell of each planet is an aspect of body-consciousness.

Chapter Six

THE OUTER AND INNER WORLD OF MAN

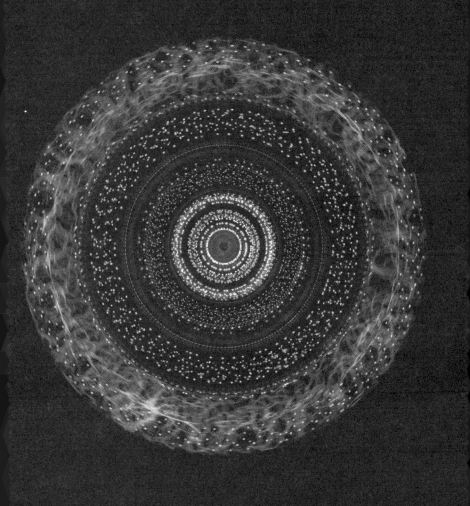

Chapter Six

THE OUTER AND INNER WORLD OF MAN

WE HAVE A SOUL, AND WE SHOULD LEARN HOW
TO USE IT. There are two levels of man: The internal soul-
consciousness or reality level and the external body-
consciousness referred to as the physical, illusionary level
elaborated upon in the previous chapter.

Prior to the sin of Adam, which was the negation of the
Lightforce, the entire universe existed and remained
connected at the level of *Ruach*,[144] unfettered by the claims of
space and time, un-shadowed by entropy and death. Body-
consciousness was completely dominated by soul-consciousness.
However, when the fall or sin of Adam occurred, he fell from
that blessed state of consciousness into *Nefesh*, where body-
consciousness could raise its ugly head, dragging the world
down with him.[145] His descendants have been struggling to
regain the Eden of his creation ever since.

There are many states of consciousness and not all of them are
benign. But because man can never separate himself from the
universe, all of these states, even the lowest, contain power
with which only fools will trifle. The soul's only defense, as it
journeys through the myriad levels that make up the
metaphysical universe, is to achieve at-one-ment (remove the
hyphens and receive yet another word of power) with the
Lightforce. One of the best road maps clarifying the way to
that goal is kabbalistic teachings.

The revelation of the *Zohar*, with its careful scrutiny of the physical and spiritual anatomy of the individual, allows us to learn exactly what exists in the realm of the galaxy, and as well, in the realm of undiscovered celestial objects. The *Zohar* states:[146]

> *Upon man rests the entire movement and strings of the universe. For there is not a member in the human body that does not have its counterpart in the universe as a whole. For as man's body is divided and subdivided into sections and all are poised upon levels of different magnetic fields and intelligence, by which each react and interact with the other so as to form one organism although remaining independent, so is the entire universe based upon parallel and different levels by which each section, each segment of the universe is related and interrelated with the other.*

Two remarkable statements emerge from the *Zohar*. Man, by kabbalistic standards, is the determining producer and director of movement in this universe. Furthermore, man is featured as a carbon copy of everything, both in the celestial and terrestrial realms of our universe. Both declarations tend to sound incredible in light of man's latest space technology, where the only place men have walked beyond the Earth is the moon.

The pure scientist may shake his head in despair at the previous *Zoharic* statement, but the *Zohar* makes it clear: the body of man is related to our entire galaxy and universe. An inherent dynamism impels the *Zoharic* conclusion that the whole of the cosmos consists of soul- and body-consciousness.

This revelation, by and within itself, is overwhelming. To attach the idea of consciousness to the cosmos is what I consider the most daring of all *Zoharic* declarations.

This concept is further extended by the Psalmist, King David, when he wrote, "The Heavens declare the glory of the Creator, the expanse of the sky tells of His handiwork. Day following day brings expression of praise, and night following night bespeaks wisdom."[147] We must also note that the positions of energy-forces should not be seen as roaming around in a random sort of way. It is our inability to define and recognize the various entities of intelligence at the subatomic or celestial level that bars us from well-defined concepts.

Rabbi Shimon Bar Yochai, who achieved an extremely high degree of an altered state of consciousness, found no difficulty in defining precisely what was taking place. He could tap information that made the absolute future and the past very distinct realities. Assuming we are now ready for the new age of infinity, how do we go about gathering this new kind of information?

The author of the *Zohar* foresaw this spark-making Aquarian period by providing some of the knowledge that now becomes necessary.

The *Zohar* describes the intimate relationship between the celestial realm of the cosmos and the terrestrial realm of man. This relationship explains our compulsive fascination with the unknown out there in our universe, our fascination that has existed for as long as we have taken notice of the structure and order of the universe. Our preoccupation with the cosmic connection arose when a fundamental awareness of the enormous and profound dimension of the Celestial Realm and its powerful influence over our affairs became firmly established.

Consequently, the assumption of a cosmic duality is a most daring and revolutionary step forward in our understanding of the power of the subconscious mind. There have been many scientific attempts in the past 300 years to explain and describe the enormously arranged and beautifully designed universe we now inhabit. However, these attempts leave many of the conclusions hanging on a thread. Scientists still cling to the belief that the understanding of our cosmos does not lie in the original initiation of its beginning structure and organization but rather upon understanding the laws and principles of nature that maintain the cosmic system and force it to operate in an orderly fashion.

To ignore the purpose behind each cause and effect, simply because this issue is very complex and subtle, is an easy way out for physics. Probability, which still plays a great role in subatomic physics, fails to cope adequately with the Celestial Realm, which lies beyond gravitational influence. Such an uncertainty might well spell the end of the road for physics as an exact science. Rabbi Shimon Bar Yochai, in his *Zohar*, realized that to face the unknowable it is necessary to probe and discover the cause of events, an undertaking that presents a difficulty with which contemporary physics cannot cope.

Kabbalists describe the cosmos in terms of positive and negative energy-intelligences. The internal non-material consciousness of celestial bodies, are, like the soul of man, in direct communication with the Lightforce. The seven planets act as orbital channels for the seven diffused forms and levels of cosmic intelligence known as the Seven *Sefirot*.[148]

From a kabbalistic world view of our universe, we merely reflect inward, no less than the physicist probing further inward towards the elusive world of subatomic phenomena. Rather than take the sun, for example, as the initial primal force of energy, the *Zoharic* point of view states that all manifested forms, or energy, are merely the result of channels of natural, intelligent cosmic emissions of the Lightforce. The Lightforce becomes enclothed within orbital vessels or channels. These cosmic beacons all originate in the Lightforce.[149] The cosmos becomes varied as it passes through these diversified terrestrial channels.

It comes as no surprise when these cosmic intelligences are discovered, and seen as messengers from the Lightforce, seeding Earth with complex organic intelligent forms. These encapsulated intelligent coded messages, known as *Sefirot* or metaphysical DNA, are the primal forces that account for our grand solar system and for the Earth's cosmic division. One day, some of mankind's most talented minds may decode these interstellar messages and discover the intelligent civilizations of outer space, civilizations that are, and have been known, by the kabbalist. In a word, these extraterrestrial messages are the complex forms of *Sefirot* that encapsulate the Lightforce.

What appears to emerge from the *Zohar's* revelation is that forms of intelligence that emanate from these *Sefirot* are directly responsible for our universal manifestation. More importantly, they are the forces that impel our day-to-day activities. These advanced extraterrestrial non-corporeal beings, in a solar system similar to ours, direct the orbiting structures of our own universe and subsequently display the varied quantified degrees of the Lightforce. All life dances to the music of astral influences as more clearly demonstrated by cyclical phenomena.

Unseen extraterrestrial forces affect and decidedly determine the ups and downs of many phenomena related to mankind. By tracking the cycles of ups and downs of terrestrial life and by identifying these cycles, the curious fluctuations clearly suggest a metaphysical pattern very closely resembling what DNA is to us.

The planets and signs of the zodiac mark their celestial imprint on the face of Earth. This might be compared to the soul of man, which makes one individual distinctively different from another. The soul is responsible for a person's creativity, free-will and emotions, such as love, hate, fear, and warring instincts. The seven, non-material entities (*Sefirot*) cause the individual diversity of the universe and Earth.

However, the Seven *Sefirot*, the heart and soul of each planet, are encased within a corporeal, physical shell observed by Earth's inhabitants. The knowable, observable shell of each planet is, according to the *Zohar*, an aspect of body-consciousness. As such, the usual limitations pertaining to the body-consciousness of man applies also to the celestial realm. Body-consciousness, whether it exists within mankind or the

Celestial Realm, causes layer upon layer of negativity to be built up around the person. The greater the Desire to Receive for Oneself Alone, the blinder we become to the Lightforce.

The illusion of darkness brought on by body-consciousness is the cause of problems and difficulties. Placed in a position or condition of vulnerability, mankind then robotically accesses into the body-consciousness of celestial influence. When negative influences of the Celestial region reign over the universe, mankind, without the benefit of a security shield, becomes inundated with chaos, disorder, and misfortune. At times, even his health may be in jeopardy. If, for example, the Desire to Receive for Oneself Alone overtakes a person during the zodiacal domination of Cancer, that individual has accessed into the disease of cancer.

From a kabbalistic point of view, the dreaded disease of cancer has its origin and beginning during the reign of the Cancer sign of the zodiac. It is by no accident that Abraham the Patriarch designated the fourth sign of the zodiac by the name of Cancer. This was done with the express purpose of sharing information for readers of the first publication on the Kabbalah, the *Sefer Yetzirah*, Book of Formation.[150] If naming were an insignificant aspect of the entity being named, then Abraham could have specified any number of names for this sign of the zodiac.[151]

Furthermore, the sign of the crab, which speaks out during the Hebrew month of *Tammuz* (Cancer), would not have the deep significance it does if there were nothing in a name. The purpose of the crab's appearance in the heavenly constellation was to provide an insight into this month's body-consciousness. As mentioned previously, "The Heavens declare the glory of the Creator."

A good starting point for our investigation into the body-consciousness of the crab is its mode of locomotion. In walking or crawling, most members of the crab genus display the peculiar characteristic of a sidelong gait—an unusual procedure for getting from one place to another. However, looking more closely at this peculiarity, we arrive at some interesting considerations. There are some marked differences between walking forward and walking sideways. The distance covered is far greater in a forward motion. In walking sideways, only one foot or side can be in motion at a time and continuous activity does not occur in a sidelong journey.

What does this distinctive characteristic imply? How and why does the Cancer sign of the zodiac relate to the crab? The Psalmist made the point with his remark that, "The Heavens declare the glory of the Creator" and implied that the signs of the zodiac reveal the internal energy-intelligence of a particular month. The cause and underlying reason for cancer is bound up and correlated with the idea of fragmentation and discontinuity, the trademark of the Desire to Receive Oneself Alone.

Crawling sideways prevents continuous motion as opposed to a forward movement. Prior to the left side commencing its activity, the right side must bring its movement to a complete halt. Try walking sideways towards a left direction. Before we can lift our left foot in motion, our right foot must rest alongside the left foot and for a moment our body must come to a standstill. When we walk in a forward position, either foot is always moving.

Soul or internal consciousness exists as a continuous, eternal circuitry of energy, without any interruption in its energy-intelligence flow. Body-consciousness represents and symbolizes a constant illusion of an interrupted, fragmented flow of energy-intelligence that results in chaos and disorder. Consequently, the crab portrays the symbol of cancer, which, in essence, characterizes short circuitry and interruption of an energy flow.

The domination of a Desire to Receive for Oneself Alone is the nature of the crab. Therefore, its internal essence of body-consciousness determined its physically manifested movement and the sign of its direction. The essential problem of the disease of cancer and the havoc it reigns upon us stems from the fact that the energy-intelligence of a Desire to Receive for Oneself Alone penetrated and invaded the wholeness and unity of the individual.

The cells of a human body are created as, and connected with, an all-embracing unified whole of soul-consciousness. During the course of one's lifetime, if an individual becomes vulnerable to the lack of fulfillment and short circuitry of the Desire to Receive for Oneself Alone during the month of Cancer, he or she becomes a victim to the scourge of cancer. The loss of a loved one or other misfortunes thrust the individual into a state of depression, meaning that lack of fulfillment overcomes the person. This condition creates an affinity dominated entirely by a body-consciousness. When one is totally consumed in matter of the flesh, illness is given permission to enter.

Inasmuch as the zodiac sign of Cancer reigns supreme during the Hebrew month of *Tammuz*, persons who succumb to a state of unhappiness or lack of fulfillment are immediately drawn to the body-consciousness of Celestial Cancer influence. The origin or beginning of chaos takes place during this month of Cancer. It is, therefore, essential that we take great care in avoiding the pitfalls of depression, unhappiness, and similar mental states of lack in the month of the crab. Connecting with the crab's body-consciousness leaves us open to its dangerous cosmic influence.

The point that I am making is that Celestial influence combines a dual form of energy-intelligence. The positive or soul-consciousness of Celestial entities regulate and determine the brighter, happier moments of our existence. Body-consciousness, on the other hand, represents the darker side of existence, invoking and exerting the influence of chaos, disease, and misfortune into our lives. Society then suffers the trauma of disorder and destruction, and seeks an illusionary, temporary program that might alleviate the pressures brought on by causes that seem to be indeterminable. In reality, the culprit, and underlying cause, lies within the body-consciousness of the Celestial region, and not within the terrestrial plane.

Solutions to a problem that do not include an accurate precise description or understanding of its origin are not viable solutions. Symptomatic reasoning simply ignores the true and basic causes for the overwhelming enigmatic behavior of mankind. We must note that very little has changed in the area of social progress. We might do well to turn in other directions for an understanding of what makes people behave the way they do.

The Kabbalah technique can and does support man in his quest for a life replete with satisfaction and blessing. To insure fulfillment, we must find ourselves unified with the soul-consciousness of the *whole* of the cosmos. The concept of soul-consciousness extends beyond the Celestial region of the planets and zodiac signs of our constellation.

We should not be surprised if computers and other high tech mechanics begin to exhibit their own consciousness. The designers who put in the programs, and the consumers who use these machines, all leave their mark on these contraptions. The closer we come to subatomic particles and atoms, the closer we are drawn to the psychology and robotic consciousness of these instruments of communication. These instruments have no free-will, by and within themselves. They do, however, have an internal consciousness to perform as they have been instructed.

The material, corporeal aspect of these information machines fall within the category of body-consciousness. As such, they become subject to, and are influenced by, negative environment and association. The table in a restaurant formerly occupied by negative occupants should be avoided. Their vibrations continue their negative influence by means of the body-consciousness of the chairs and tables. If you have not enjoyed your meal at one of your favorite restaurants, it may well be the fault of the table and not the chef!

Becoming attuned to the environment is as essential as becoming familiar with the kind of food we eat or the company we keep. Before moving into a new apartment or a new home, we must be fully informed as to its former occupants. The vibrations continue to make their presence felt long after the occupants have left. If misfortunes or grief had been the trademark of the former inhabitants, then the new homeowners can expect this negative energy-intelligence to continue and have an influence on their lives and relationships.

The idea of former occupant vibrations extending their influence on its new occupants has been elaborated upon by the *Zohar* [152] and dealt with in the biblical section of Leviticus.

> *Rabbi Yitzchak was going to his father's home when he saw a man turning aside from the road with a load tied to his shoulder. He said to him, what is the bundle that you have on your shoulders? But he gave him no answer. So he followed him till he saw him entering a cave. He went in after him and saw a cloud ascend from the ground, while a man went into a hole and he lost sight of him. Rabbi Yitzchak left the cave in great fear.*
>
> *As he was sitting there, Rabbi Yehuda and Rabbi Chizkiyah passed by. He went up to them and told*

them what happened. Rabbi Yehuda said, "Thank the Lord for delivering you. That cave is where the lepers of Sarunya are. All the inhabitants of that town are magicians and go into the desert to get black snakes, ten years old or more, and by not being careful with them, they become lepers. All their magic arts are in that cave.

On their way, they came across a man who had with him a sick child tied to an ass. They asked him why the boy was tied. He replied, "I live in a town of the Arameans. My son used to study Torah every day. Three years I lived in the same house and observed nothing wrong. One day my son went into this house to repeat his lesson, an evil spirit passed before him and distorted his eyes, his mouth, and hands. He was not able to speak. So I am going to the cave of lepers Sarunya to ask whether they could show me how to cure him."

Rabbi Yehuda said to him, "Do you know of any other person having come to harm in that house before?" He replied, "I know that a long time ago a man did come to harm there but most thought it was an illness, some said it was from the evil spirit of the house. However, since then many persons have been in the house without suffering any harm." They said: "This proves the truth of what the companions said, 'Woe to those who disregard their words.'" Rabbi Yehuda said, "Woe to him that builds his house by unrighteousness."[153] *Because wherever there is righteousness, all evil spirits fly away from it.*

Yet whichever comes first to the place, takes possession of it. Said Rabbi Chizkiyah, "If so, the Holy Name is on the same level as unclean spirits?" He replied, "It is not so, if the Holy Spirit is there first, no evil spirits can be seen there, much less approach. But if the evil spirit is there first, the Holy Name does not rest upon it.

… The best thing is to leave the house. But if this cannot be done, it should be rebuilt with fresh wood and stones and a little away from the previous spot, with the mention of the Holy Name."

What seems to emerge from this *Zohar* is that inanimate objects are really not as lifeless as we once thought. The air, earth, and rocks are made of vibrating molecules and atoms. They consists of particles that interact with one another by creating and destroying other particles. The atoms of the elements and our body participate collectively in a cosmic dance of energy and activity. Things out there are not independent of ourselves. Everyone and everything acts and interacts upon each other.

Consequently, the varied soul-and body-consciousness of our universe combines to create and influence each other's activities and movements. There appears to be little room for mankind to act or behave in a manner that might provide for free choice and determination. This concept, in fact, lies behind the quantum theory and leads some scientists to conclude that man lives in a free universe where no one can actually be held responsible for his actions. After all, because outside influence is so massive and intense, what degree of free-will does mankind have?

Stretching this idea to its extreme, in his defense, a murderer or assassin might claim that forces beyond his control were responsible for his criminal activities. For the present, we might consider ourselves fortunate that scientific knowledge and information have not become part of the public domain.

Despite these conclusions, the Revelation on Mount Sinai does indicate room for free-will. The Ten Commandments, which include the prohibitions of criminal activity, clearly state a degree of control in mankind's decision-making process. While on the one hand, we are continually bombarded by an infinite array of thoughts, the kabbalistic world view is that we can and do exercise free choice.[154]

All signposts seem to point in the same direction. Outside influences contribute a great deal toward our behavior and manner of doing things. There is no way in which we can just make these energy-intelligences fade away. They are part of our universal landscape. They will either assist and support our objectives or create a sort of chaotic environment where things begin to go awry. Tapping the positive energy-intelligence of our cosmos or creating security shields to protect us from cosmic negative activity is what the study of Kabbalah is all about.

Because we are acutely aware of our limitations and the inherent prejudices into which we are born, our collective consciousness affords us very little opportunity to achieve a quantum positive attitude. When we scan the *Zohar* and participate in Kabbalistic Meditation, we foster a more cosmically positive level of energy-intelligence for ourselves, the world, and the universe.

While deliberately avoiding preconceived ideas or political links, Kabbalah endeavors to excite the individual's awareness and recognition of mankind's greater potential. The Kabbalah technique will both permit and induce all of humanity to realize that *whatever serves the collective consciousness of humanity also serves the needs of the individual.*

There is another important matter about which much has been written, and yet, I feel very little clarity has emerged. I am referring to the subject of the subconscious mind and its power.

Man's mind and his mental processes have always seemed as mysterious and fascinating as the universe itself. Only in recent times has the investigation of the nature of the mind become the domain of experimental science. However, the more light that is cast on the nature of our mental processes, the more questions are raised.

Intelligence is, by far, one of the most desirable of all human characteristics but *intelligence* does not mean the same thing to everyone. Some claim it refers to an ability to perform mental functions successfully. But we ask, "Which functions are to be considered relevant?" Reasoning capabilities have something to do with intelligence; memory plays some part within the scope of intelligence.

What about those people whose inventiveness is extraordinary but whose memory fails just when that piece of vital information is required? What about those people who supposedly belong at the other end of the totem pole, the idiot savant? This term applies to an individual who finds great difficulty in performing important everyday mental functions, while at the same time displays some extraordinarily brilliant capabilities.

To this very day, intelligence tests have frustrated many psychiatrists and analysts. The intelligence quotient, also known as IQ, has come under necessary constant revision simply because the everyday stimuli upon which the tests rely are constantly changing. This has caused great disappointment in finding the means by which to measure intellectual capability.

IQ testing was designed in the 1920s, principally for children and young people. These tests are not very useful or successful in testing adults. The reason for their limited success with children is the limited scope of exposure to life that most children experience. While intelligence tests are tremendously useful, and will doubtless be with us for a long time, nevertheless mental health experts are still not sure what these yardsticks actually measure.

Until approximately 20 years ago, most psychologists were of the belief that intelligence was governed by heredity and was consequently invariable throughout one's life. To the extent that IQ tests measure intelligence, there are strong indications that intelligence can actually change during an individual's lifetime. As more and more people were being tested, it became increasingly difficult to ignore the overwhelming evidence that intelligence could rise, with education being one of the contributing factors.

With the discovery of this evidence, the age old question again comes to our minds: Then what is intelligence? Let us face up to the naked truth that for all the scientific research and data that have been accumulated over the years, we are still not any closer to a clear-cut definition of what constitutes intelligence. No test yet devised can really measure naked intelligence, intelligence entirely devoid of, and not influenced by, the effects of experience of the learning process. A better than average performance may indicate a higher than average experience. Good study habits, together with an environment generally conducive to the enhancement of intellectual skills, frequently will be reflected in a child's score on an intelligence test.

On the other hand, the kabbalistic view of intelligence is completely relative to, and dependent upon, each unique and particular individual. From a kabbalistic standpoint, there are essentially no yardsticks by which to measure the intelligence of a person. The more comprehensive measurements, which unfortunately are concealed, relate to one's prior incarnation. However, as a result of the *tikkun* process,[155] an individual, irrespective of social environment, family or other involvements, can get started on a spiral of intellectual growth.

The levels of intelligence are known by their coded kabbalistic names in which the kabbalist can recognize the specific level of intelligence. The *Sefirot*, or levels of consciousness, fall into five gradations—*Keter, Chochmah, Binah, Zeir Anpin,* and *Malchut*—towards which mankind can aspire and achieve higher levels of awareness and creativity.[156] From a kabbalistic viewpoint, there is no such thing as the power of the subconscious mind. The mind of each person provides the same ability and capabilities. The mind is merely a channel.

The difference in the use of the mind lies merely and solely within the "desire" of each of us to grow spiritually.[157] To acquire the knowledge and understanding by which to tap these awesome levels of cosmic consciousness is most definitely a beneficial pursuit. Psychiatrists and psychologists have mistaken the subconscious mind for the consciousness, which already exists in the cosmos. The mind is merely the channel by which we connect to the varied levels of consciousness that belong to the Upper Realm of the cosmos.

From a kabbalistic point of view, there is no hidden power of our subconscious mind. Many spiritually-orientated people around the world have already begun their journey in acquiring the infinite intelligence that all of us can possess. By spiritually-orientated, I mean those people who have come to the realization that they must strive towards a sharing consciousness and introduce the principle of Restriction into their lives on a daily basis.

The infinite intelligence of the cosmos, which is not subject to time, space, and motion, can reveal everything we need to know at every moment in time, provided we are *open-minded* and *receptive*. Though invisible, its forces are powerful. We can draw upon the awesome power of the cosmos at will, increasing our consciousness and awareness beyond our wildest imaginations. We can bring into our life more power and wealth, more joy and happiness, and most importantly, greater health, by learning how to tap and reveal the hidden power of the cosmos through mind channeling.

This vast reservoir of cosmic beneficence belongs to all of us. However, only those who gain the necessary knowledge and understanding to expand the capacity and ability of their mind power can be the recipients of this infinite source of wisdom. It is our right and privilege to discover this inner world of consciousness. Though invisible, its forces are awesome and mighty. They can enable us to find the solution for every problem, and more importantly, the original cause of every effect.

Intrinsically there is but one mind. However, the mind possesses two functional sets of consciousness. The two functions are directly connected with the kabbalistic doctrines of Inner Light consciousness and Encircling Light consciousness.[158]

Before explaining the doctrine of Encircling Light consciousness, I am devoting approximately 20 pages to the kabbalistic doctrine of Inner Light consciousness, which embodies the concepts of unconscious or subconscious activity, as well as the mind, brain, sleep, dreams, immune system, IQ intelligence, and the *tikkun* process.

The Inner Light consciousness consists of the rational conscious mind and the unconscious non-rational mind— both expressed by the individual. Conscious awareness is an automatic process of the physical brain. Although scientists claim there is a relationship between consciousness and the brain, they really do not know what that relationship is. For the present, what might be considered as mind-consciousness falls within the category of Inner Light consciousness.

The unconscious or subconscious mental activity of the mind is considered to be the invisible fabric of mind. Medical scientists have long skirted the idea of any intellect below conscious awareness. Psychosomatic or emotional problems, or persons worrying without reason, are some of the behavioral results of abnormal mental activity that produce stressful subconscious struggles. The role of the unconscious in emotional disorders is rarely mentioned by psychiatrists, despite the mind's unique ability to form abstract concepts or images that can actually create physical change in the body.

The conscious and unconscious realities are unique and individualistic. They are established within each person at that level of intelligence, which depends entirely upon the reincarnation dictates of previous lifetimes. Therefore, to a certain extent, these levels can be measured. Rarely do these individualistic conscious and subconscious levels of activity change or expand.

Inner Light consciousness is subject to the personal *tikkun* of the person. As such, our total conscious awareness, together with our subconscious, is limited, possessed by the fragmentation of time, space, and motion. The origins of human behavior can be traced to prior incarnations and are *not* affected by the different circumstances of human environments.

The mind is similar to a computer that is programmed in a particular way and will then process information accordingly. Prior incarnations determine the particular program for our mind. Environmental stimulation or observations will be processed by our programmed mental computer. The mind is nonmaterial, a unique intangible product of our prior lifetimes that processes the information fed to it by our surroundings. After properly digesting the material, the mind reaches conclusions. The mind then releases its findings to the brain.

In a way the mind possesses the ability to regulate and control the brain. The brain, as a product of mind activity, controls the very functions of the electrical and chemical nature of the entire body. The brain, according to the kabbalistic teachings, is the *Keter* (Crown) or seed of all physical manifestation and activity.[159] Consequently, after the egg is fertilized and the baby grows in size, it changes in proportion and becomes more recognizably human. The head appears first, followed by the appearance of limb buds.

The kabbalist always seeks first causes. The head is the *Keter* of human development. Its internal energy-intelligence consists of the *Sefirot Keter*, which has the awesome power of all-inclusiveness. Similar to the seed of a tree,[160] which includes and encompasses all future physical manifestations, the brain thus possesses the awesome power to control and regulate the distribution of its mind power produced by the mental, mind computer printout.

It is beyond our ability to comprehend the awesome power of the mind as it directs the electrochemical processes of the human body with such precision. In some way the mind establishes goals to accomplish a specific assignment. By controlling particular physiological activities, the mind even projects the results of what is to be achieved in the process. Indeed, according to kabbalistic teachings, the task to be accomplished has already been established long before the elements of the brain are called upon to perform the functions necessary to carry out the mind's intention.

The idea that has been grasped by the mind *already* includes what is needed to be done to accomplish its objective, as well as the intentions and decisions to control the required neurophysiological processes needed to achieve its goal. To appreciate this extraordinary power of the mind, we have only to consider the enormous amount of advance information provided by the mind's former incarnations. Because of the mind's built-in "robotic consciousness," flaws or mistakes in relaying instructions to the brain are non-existent. The brain's function is one of executing the intentions and decisions of the mind at the physical level.

Abstract thought, advanced reasoning, learning, judgment, and planning—all of these would be impossible without the highly developed human mind. But the human mind is much more than an energy center of intellectual activity. The mind regulates, directs, and coordinates all the sensory impressions we receive, all the emotions we feel. Because of our individualistic *tikkun* process, which is bound up with our own prior lifetimes, we can have an insight as to why each of us views the same things somewhat differently. We can understand why we react diversely to the same circumstances. In short, our individual *tikkun* process is what differentiates one human being from another.

I have compared the mind and brain to the computer. Just as computers depend on instructions to do this first and that second, the mind-brain follows the instructions of the *tikkun* process program. But it is *there* that all comparison with the computer stops. While a computer processes information a single step at a time, the brain, with its trillions of cross-linked neural connections, processes information along millions of multidirectional pathways all at the same time. A computer cannot decide that it is wasting its talents or that it should embark upon a new way of life. A computer cannot dramatically alter its own program. However, a person with a human brain must reprogram himself before moving in a new direction.

The brain, the command station of the nervous system, controls our activities as well as the functioning of our internal organs. It is the computer that connects the mind and the brain. It is also the intrinsic ability of the human mind-brain to exercise free-will. The *tikkun* process permits each person to alter his or her mind program and thus bring about changes in the ultimate manifestation of the brain's newly acquired program.

Compared to the other organs of the body, the brain is the most important. One of the most remarkable phenomena concerning the brain is its insensitivity to pain. Brain tissue rarely hurts, even when a surgeon's knife cuts into it. The brain, however, is subject to what is known as "referred pain," pain that arises in one part of the body causing the surface of the head to hurt. What this phenomenon indicates is that the brain belongs to our consciousness and the world around us. The brain is the citadel of the human spirit. The brain, while a physical organ of the human body, can also be seen as belonging to the unseen, metaphysical world around us.

The brain awaits its instructions from the mental *tikkun*-processed computer, which apparently remains forever concealed from the physical brain. To this day, most experts acknowledge that they do not yet understand everything about the mind. This mystery will continue to prevail. The mind and its link to the human body, as well as to the brain remain, of necessity, a mystery because the mind is metaphysical and directs its printout through the crown of the human body, the brain.

The great gap that science must bridge in the knowledge of the mind remains as follows: "How are the actions of the nervous system translated into consciousness?" "What role does the brain have in this scenario?" As new knowledge is discovered, science is making amendments and replacing prior points of view. The result is the creation of more and more complicated questions, while still leaving the key question unanswered, "What is the mind?"

Rabbi Isaac Luria (the Ari), with the stroke of his pen, laid to rest the mysteries concerning the complex anatomical structure of the brain.

> *The brain is referred to as Da'at (Knowledge) for the Sefira Keter (Crown) is bound up and connected with the brain.*[161] *Keter is the link between the Lightforce and the brain, for it is an essence that is inclusive of both.*[162]

What seems to emerge for the Ari is that the internal energy-intelligence of the brain, known by its code name *Keter* (Crown), has the ability to connect with metaphysical intelligence, in this case with our particular mental computer. The second quality of the brain is its all inclusiveness with the entire body. Hence the brain's ability to control and regulate the entire human physical structure, which for a king is symbolized by the crown. At times, the crown connects to a physical body, thus crowning the individual as king. At other times the crown is the symbol or metaphysical representation of what a crown portrays.

Thus the brain serves a two-fold purpose. The first is to channel the printout of its mental computer. The second is to make manifest the instructions and regulate its dynamic movement within our physical, corporeal body. Yet scientists, as well as the philosopher René Descartes, made a complete and total division between mind and body. Thus, once the inclusive nature of the mind was admitted, a host of new problems arose, many of which remain unanswered to this very day. If the mind is accepted as the knower of all the enormously complex matters and relationships of life, "How does the mind *know* this and relate to it?"

The brain was, and is still seen as a warehouse stuffed with all kinds of furniture. All of it—intuition, *déjà vu*, flashes of creativity, everything we know about the world around us—has to be accounted for. The hardheaded realists, known as the empiricists, refuse to acknowledge the idea of innate ideas. They refuse to accept the notion of a brain born with a *small* basic supply of furniture—the Inner Light—despite the overwhelming evidence that supports this idea. Computers, designed to imitate the human thought process as far as possible, remain oblivious to the dark world of the unconscious. The mysterious realm of the unconscious is, kabbalistically speaking, the mind or our mental computer designed by our former lifetimes and prior incarnations.

The complex machinations of the brain provide a glimpse of our own metaphysical universal mental computer. To appreciate the near infinity and vastness of the mind, which will eternally remain concealed from any physical mode of detection, let us briefly examine the awesome complexity of our nervous system. Complete networks of nerve cells run throughout the body, connecting *every* distant bit of tissue with the more than 10 billion nerve cells of the governing brain. Lightforce or electrical impulses travel along these superhighways connecting the infrastructure at speeds of up to 250 miles per hour. They perform incredible leaping feats across narrow gaps between cells. The communication system far outperforms any high tech telecommunication system designed by man. The various networks simultaneously perform a dazzling array of tasks.

Whatever the nature of the mind or our mental computer, the mechanisms through which it expresses itself are beyond scientific belief. Consequently, when considering the mind itself, where, and from which point, can scientists begin their investigation or research? Scientists face insurmountable problems and obstacles when dealing with the expressionable brain. Therefore, examination of the mental process or how the mind functions, is a study that has no starting point. The increasing dilemma faced by psychiatrists and other researchers continues unabated. Today, there is hardly a respectable neuroscientist who thinks the mind exists apart from the functions of the physical brain and body. Yet what some researchers have called the "ghost in the machine" continues to haunt their efforts to scientifically describe human conscious and unconscious thinking. The answers lie unmistakably in the realm of metaphysics, the reality level where the biochemical model comes to an abrupt end.

Another aspect of human existence that supports the kabbalistic theory of the Inner Light phenomenon is that of sleep. On average, we spend almost a third of our lives in sleep. Yet, we know little about precisely what sleep is supposed to accomplish.

Some researchers believe sleep serves some restorative purpose. Our bodies require sleep but how and why does sleep provide these necessities? Why do we awake refreshed? To this day, the answers remain unclear. Science knows little of the constant night life of the brain. Only in the 1950s, did investigators at the University of Chicago find that sleepers periodically make rapid eye movements. When these subjects were awakened during such movements, they testified that they had been dreaming.

Researchers also determined that the heartbeat quickens during dreams and the brain-wave pattern becomes similar to that of someone alert. Quite an active night life for the brain whose enormous and complex activity never permits itself the luxury to enjoy a vacation or a day's rest. With all this activity, combined with the many chores the brain has to regulate, control and initiate, all at the same time, it is little wonder that our mental institutions are filled to capacity. In fact, with the mind being taxed as it is day and night, it is a miracle that all of us have not wound up in a mental institution.

The sleep phenomenon, combined with all the exciting mysteries that surround it, establishes conclusively that there indeed exists a "ghost in the brain machine." The ghost is our Inner Light, the mind, our mental computer. There is *no other* rationale that can explain the awesome ability of the brain to perform 24 hours a day for years on end. While there are many similarities in the way electronic machines and the brain function, no one has come forward with a claim that a man-made machine will someday equal the performance of the brain.

Both the brain and the computer process incoming raw material with the support of complex circuitry systems. Both demonstrate built-in systems for storing enormous quantities of information in their memory banks. However, the brain calls upon stored-up information that it has not experienced before—at least not in this lifetime. The infinite brain is unequalled. The computer, on the other hand, can only access into stored information placed there by a programmer or program.

The subject of dreams and sleep requires a complete and massive investigation, which is dealt with extensively in the study of Kabbalah. The purpose of this chapter is solely to present a solid case for the existence of a mental computer, completely independent of, and not subject to, the intellectual activity of the brain. While the brain regulates and coordinates all our voluntary and involuntary movements, nevertheless the physical manifestations expressed by our brain reflect merely our robotic consciousness. It is our mental center, the mind that initiates all activities, which are then expressed by our brain printout.

Prolonged lack of sleep results in our inability to function as usual. We suddenly find ourselves in a condition whereby we have difficulty carrying out mental and physical chores. Sleep deprivation experiments reveal that a person may experience extreme vulnerability. He or she may even hallucinate and demonstrate other signs of mental illness. Sleep deprivation is the principal method used by some cults and governments to accomplish "brainwashing."

Why do people who have trouble sleeping suffer a multitude of emotional and physical ailments, in addition to always feeling fatigued, despite their lying in bed and resting for days on end? In time, sleeping pills make insomnia worse rather than better. There is no such thing as pills that foster normal sleep.

But now let us return to our original question, "Why is sleep necessary?" This question has already been raised by the *Zohar*. Sleep will greatly benefit and improve an individual's physical and mental well-being and achieve Kabbalah's essential objective. My purpose in drawing attention to the following *Zohar* is to indicate the necessity of an awareness and knowledge of what *makes* our brain "tick." The soul is the total encapsulation of the original Inner Light. *The mind is our soul.* According to the *Zohar*:[163]

> *... the soul mounts up, returning to its source that fits it above, while the body is as still as a stone, thus reverting to its source of origin. While in that state, the body is beset by the influences of the dark side (negative body-consciousness), with the result that its hands become defiled and remain so until they are washed in the morning, as explained elsewhere. There all the souls are absorbed within the all-embracing unified whole, the Life-force... The souls then re-emerge, that is to say, they are born anew; each soul being as fresh and new as at its former birth. This is the secret meaning of the words, "they are new every morning... great is Thy thoughtfulness."*[164]

The verse above, Lamentations 3:23, means that the souls are new every morning and the words, "great is thy thoughtfulness," refers to the Lightforce's desire to absorb them and then let them out as newly born.

The soul, states the *Zohar*, requires a pause that refreshes. Following a daily battle with the body, its adversary, the soul requires an infusion of energy to continue its struggle with body-consciousness until life's end. Soul-consciousness is one of a Desire to Receive for the Sake of Sharing. Body-consciousness pursues the indulgence of a Desire to Receive for Oneself Alone. The battle goes on with both sides paying a heavy price. The soul sees itself as the enemy of the body.

The aim of body-consciousness, with respect to the soul, is to prevent the soul from achieving its *tikkun* or correction.[165] However, the body does provide the soul with an opportunity for the removal of Bread of Shame.[166] Without a body-consciousness, the idea of free-will is not a reality. The creation of the observable physical world, including space and time, permitted man to achieve the objective of a corporeal expression of the Desire to Receive for Oneself Alone.

Sleep then is a necessary function for the soul, the need for it is compelling. The soul, after a day's encounter with the body, is totally washed out. Were it not for the Creator's arrangement that our body lose its "war-like" consciousness and fall asleep, mankind would cease to function. By falling into a natural state of suspension of conscious activity, the soul now has the opportunity to return to its origin for its recharge.

Body-consciousness is like a leech, searching and snatching whatever energy it can from its enemy, the soul. If we try to stay awake for 48 hours, chances are we cannot do it. The soul, the Life-force of our body, needs energy as long as it is connected to the body. The soul's only reprieve is for the body to let it go, to permit it—as beautifully stated in the *Zohar*—to become "absorbed within the all-embracing unified whole, the Life-force."

Battered old mattresses or noisy neighbors are frequently blamed for those suffering from insomnia, among a host of other causes. The *Zohar* appears to place the blame at the doorstep of our body-consciousness. We may be permitting our bodies too many of the luxuries it so desperately craves, namely its Desire to Receive. The blessing of being asleep is that the body-consciousness is no longer aware of its being.

For this very reason I have indicated throughout this book that the physical body and body-consciousness are two distinct concepts. The former does not require rest or sleep. Evidence of this lies in the 24-hour activity of our body. The heart performs throughout the day or Heaven help us if it did not. The brain is just as active while asleep as when we are awake, unlike the popular idea of sleep as a quiet state. Our brain, even when asleep, consumes 20 percent of the oxygen that our body takes in. Our brain, whether we are awake or asleep, is always functioning. It does not move, nor does it need to rest.

Body-consciousness, mankind's Desire to Receive for Oneself Alone, "lies still as a stone" while asleep. This arrangement was necessary to permit the soul to rejuvenate itself after its weary daily struggle with our body-consciousness. Therefore, the whole idea that surrounds the phenomenon of sleep centers on the ability to put to sleep the internal energy-intelligence of our body-consciousness.

The culprit and initial cause for *all* of mankind's physical handicaps and illnesses is bound up and connected with the rampant ravages of the body-consciousness. The body itself was created with inner healing powers that are well known in medical research. An immune system has a very impressive arsenal of armaments to ward off disease. Scavenger white blood cells, as well as other substances in our blood, called antibodies, are the protectors of our immune system. They are capable of destroying harmful invaders. The skin acts as a barrier to infection, while acids in the stomach kill bacteria.

Everyone is born with an immune system. If the body's resistance is strong enough, the antibodies eventually overwhelm the invaders, the enemies and the disease. One of the major problems facing modern medicine is that medical therapies act in place of or as a substitute for the body's natural healing powers. Antibiotics indiscriminately kill healthy cells as well as diseased cells. Drugs, such as insulin and cortisone, adjust the body's chemical levels instead of the body's system of homeostasis.

Modern medicine has undertaken to provide its own system of healing, by-passing the natural healing powers that belong inherently to the body. However, let us never forget that the conscious mind of even the finest, most dedicated medical practitioner can never think or act with the same sophistication and sensitivity of our own natural immune system. When our medical needs require a particular chemical, our inner healing system provides us with the precise measurement, where and when we need it.

But there is an even more serious problem with the biomedical approach. We have accepted the idea that the body readily accepts these chemical intrusions. Yet at the same time, there is hardly a chemical medication that does not contain a warning of side effects. In actual fact, the body vigorously resists this invasion. Chemotherapy provokes resistance in cancer cells, which makes them more cancerous. Antibiotics strengthen bacteria while weakening our immune system.

I do not claim that all drug therapy is wrong. However, I am suggesting what the *Zohar* implies, namely that the origins and cause of degeneration can be traced to the internal energy-intelligence of body-consciousness. Acute and traumatic conditions of a life-threatening nature can be attributed to the prevailing energy-intelligence of body-consciousness and can be corrected in two ways. Firstly, by reshaping our attitudes to one of a more positive nature, and secondly, by connecting with the positive energy-intelligence of our soul-consciousness to promote the restoration of the immune system.

Although strengthening and restoring of our natural, inner healing powers is included within the kabbalistic teachings, nevertheless we must still pay attention to the principles of healthful living.

One of the mysteries of the immune system is a peculiar phenomenon linked to the vital role of the thymus gland. The thymus—two oval shaped lobes—appear in early infancy behind the infant's breastbone. The thymus is responsible for the development of the immune system. During the maturation period, the immune system of the infant is supplemented by factors acquired from the mother's milk, whose internal energy-intelligence is dominated by positive essence. This positive essence serves as a security shield to prevent any form of spoilage regardless of the changes that milk products may undergo. If milk should spoil, it may still be transformed to a useful product, such as cheese or yogurt. It is interesting to note that this is not true of meat products. We know that once meat has gone bad, it must be discarded.[167]

After birth, the thymus gland produces cells called lymphocytes that recognize and protect the body's own tissues, while at the same time initiating an immune response to attack by disease. Strangely at puberty, approximately 13 years of age for boys and 12 years of age for girls, the role of producing these lymphocytes is transferred to the lymph nodes, spleen, and bone marrow. Why, and why at this age?

Some insight into this strange phenomenon, especially in light of the growing interest in the importance of the immune system, may be found in the following *Zohar*.[168]

> *And Jacob sent messengers...*[169] *Rabbi Yehuda discoursed on the following text: "For He shall give His angels charge over you, to keep you in all your ways."*[170] *According to the companions, the moment a child is born, the Yetzer haRa (Evil Inclination or the embodiment of negative energy-intelligence) attaches itself to the child to drain away the Life-force, as it is written, "sin crouches at the door."*[171] *"...at the door" means at the opening of the uterus, which accompanies the birth of a child. The term*

"sin," being a designation and (code name) of the evil energy-intelligence, who was also called "sin" by King David in the verse: "...and my sin is ever before me."[172] *He is so called because he makes man to sin every day before the Lord, never leaving him from the day of his birth till the end of his life. The Yetzer haTov (Good Inclination or positive energy-intelligence) first comes to man only on the day he begins to purify himself (at puberty). What day is that? When he reaches his thirteenth birthday (for a girl on her twelfth birthday). From that time on, the youth finds himself attended by two companions, one on his right and one on his left, the former being good and positive, the latter evil and negative. These are the two veritable angels appointed to accompany man, continually. Now, when man exercises to do good, the evil inducer is humbled before him, and the Right gains dominion over the Left, and the two together join hands to guard the man in all his ways. Hence it is written: "For He will give his angels charge over you, to keep you in all your ways."*

This same version, with an additional design, is the subject under consideration in another section of the *Zohar*.[173]

"And Jacob dwelt in the land of his father's, sojourning in the land of Canaan."[174] *Rabbi Chiya discoursed on the verse, "Many are the evils that surround the righteous, but from them all, the Lord saves them."*[175] *Many indeed are the enemies with whom a human being has to contend from the day*

the Lord breathes a soul into him, in this world. For as soon as man emerges into the light of day, the Evil Inclination or negative energy-intelligence is at hand in readiness to join him, as is written: "sin crouches at the door,"[176] because that is when the Evil Inclination associates with him.

Come and behold, this is true. Note that the animals, from the day they are born, are able to care for themselves, and avoid fire or other similar dangers. Man, on the other hand, has a natural tendency from birth to throw himself into the fire. The reason is that evil intelligences dwell within him and from the beginning lure him into evil ways.

Scripture states: "Better is a poor and wise child than an old and foolish king who knows not how to heed a warning."[177]

The child here signifies the positive energy-intelligence, who is so called because he is, as it were, a youngster by the side of man, whom he does not join until he is at the age of thirteen years. "He is better than an old and foolish king," meaning the evil intelligence, who is called "king and ruler over the sons of men" and who is considered old since as soon as a man is born and sees the light of day he (the king-evil) attaches himself to him. And he (the person) is foolish, not knowing how to receive warning. As King Solomon says: "The fool walks in darkness."[178]

The startling revelations of the *Zohar* provide the necessary meaning and answers to the whole of Creation, including the anatomy of man.[179] We now understand why everyone is born with an intact but underdeveloped immune system that matures shortly after birth. Following the *Zohar*'s confirmation that after leaving the security haven of its mother's womb—protected by watery fluid and fed by the mother—the newborn infant is vulnerable to attack by the *Yetzer Hara*, the negative energy-intelligence that immediately attaches itself to the child. Lacking the protection previously provided by the child's mother, we now understand the vital role of the thymus gland: To provide the security shield necessary to protect the infant's immune system.

However, upon reaching the age of puberty, 12 or 13 years of age, the *Yetzer HaTov*, the positive energy-intelligence, now dwells within the child and provides the necessary security shield. Consequently, there is no longer a need for the thymus gland. Now the soul-consciousness of the individual is brought into the scenario. The health of each person will now depend on his or her activities, whether they are of a positive or negative nature.

Taken altogether, what seems to emerge from the preceding *Zohar* is the significance of, and the determination by, the metaphysical intelligent forces referred to previously as the "ghost in the brain machine." Although sleep and rest, for the most part, are not a requirement for our body-consciousness, it is essential that our soul-consciousness have a reprieve from its constant battle with the Desire to Receive for Oneself Alone. The Inner Light determines our physical and mental state of health.

Our example of the thymus gland is yet another illustration of the metaphysical energy-intelligence that governs our immune system. The kabbalistic teaching technique was designed with this objective, taking into account that treatment of the physical body, while no less important, is inadequate to bring about a wholistic model for improved health. The *Zohar* insight and interpretation of man's physical, corporeal body pinpoints more decisively the diagnosis of ailments and their subsequent solution. Once the underlying problems are identified and the causes are determined, the solutions are not hard to come by.

In summary, the Inner Light, coded by kabbalistic teachings, is the foundation of the infinite expressions of our physical body including the brain. Physical and mental disorders are determined by our Inner Light, the mind computer. But note carefully, this does not mean that we can abuse our body and mind and still feel confident that our Inner light will override our negligence when treating our acute conditions.

We have agreed that the final form of any human structure is related to the *tikkun* process in which the organism has lived in prior lifetimes. It is now necessary to consider the extent to which parents affect the development of their offspring. With modern technology, it has been possible to demonstrate that genes direct the development and function of every part of man's body. Actually all human traits result from the complex interplay of heredity and environment. Data from studies done in the early 1950s point to the pattern of heredity as playing a major role in the determination of offspring. Can this conclusion be considered contradictory to the kabbalistic teachings? The Kabbalah states that the *tikkun* process, and *not* the heredity factor, expresses the dominant role for intelligence, various personality traits, as well as for resistance or susceptibility to various diseases.

The *tikkun* doctrine of reincarnation only establishes that the soul requires a proper setting for the execution of one's *tikkun* process. From a kabbalistic perspective, the particular manifestation of offspring hinges essentially upon the thoughts of the parents at the time of conception.[180]

We must then question why it is that characteristics, such as height, skin pigmentation, intelligence (however explained) and specific abilities show a marked tendency towards the concept of heredity. If, as mentioned, the mental computer, which arranges the complete printout of the human being, depends on one's particular range of prior incarnations, why do genetic studies indicate that heredity plays an important role in the determination of such characteristics?

Firstly, these traits, and others like them, always vary by imperceptible gradations over a wide range of other values. There are many different genes, and while each of them may have only a small effect, nevertheless they are all operable. It is, therefore, difficult to ascertain conclusively the impact of heredity. Secondly, these characteristics do respond readily to variations in the environment, which may conceal or alter the genetic effects. Consequently, a person exposed to the geography of the equator for a length of time may appear darker than the genes might actually indicate. Or an individual genetically disposed to obesity may, if starved, be considerably thinner than a well-nourished person genetically inclined to slimness.

The concept of *tikkun*, as seen through the eyes of the kabbalist, does not conflict with the idea of genetic transmission, if, in fact, such transmission is genetically necessary. What appears as genetic similarity may actually be the result of a soul returning after death to find the particular environment conducive to achieving the *tikkun* process. Hereditary disease, such as hemophilia which is the result of a defective gene transferred from a parent, does not conflict with a person's mind-mental computer printout. If one's *tikkun* requires this condition and environment, then the returning soul must locate conditions similar to those it left behind.

In summary, the Inner Light, the mind, never conflicts with the brain and is not the brain. The mind and the brain are to be observed as the inner and outer, metaphysical and physical, aspects of the brain. They may be compared to the soul and body of mankind, where the soul, relative to the body, is considered as internal.

What is intelligence? Can we raise our Intelligence Quotient (IQ)? Is intelligence inherited?

Though most of us would like to think that we can recognize intelligence when we meet it, there still exists the simple problem of trying to explain just what it is. In truth, to this very day, researchers have not as yet arrived at a definition that satisfies all of them. Some think of intelligence as a combination of related abilities. Insight, creativity, flexibility, and the speed with which the brain can process information comprise what most psychologists would like to think of as intelligence.

Most of the numerous intelligence tests now in use reflect the view of intelligence as multifaceted. Some tests measure one group of abilities while others consider and measure such vital factors as memory span or spatial perception. Does this sort of testing conclude that if someone cannot claim instant or moderate recall of information, then there is a deficiency in that person's IQ? I find this very difficult to accept in view of the fact that there is little consensus of opinion as to what constitutes intelligence in the first place.

To assume that intelligence, like all human traits, is a product of genes and environment is likely to confound the issues and leave us in the dark. There is no single gene known that has been positively identified as contributing to intelligence. Evidence that heredity plays a major part in establishing levels of intelligence comes only from studies of closely related people. Identical twins are more alike in their IQ than others who do not share the same genes.

From a kabbalistic perspective, we raise the question, "Why were these identical twins placed in that position or environment in the first place?" In other words, we must first review the possible causes and ascertain what *initiated* the circumstances, rather than place the emphasis and study on the outcome itself.[181]

The *conclusion* that identical twins have similar IQ's must require a reflection back to the causative factor. Results of a physical nature can never explain the cause or serve as a basis for conclusion. The Kabbalah regards any research based on identical twins that attempts to prove the link between heredity and IQ as invalid. A starting point for any investigation requires, above all, answers to the "why" of things. The first question that the researcher must consider is, "Why were these identical twins, and not some other pair of twins, brought together to these selective parents?"

The answer lies within the pattern of the reincarnation *tikkun* process, which also provides a reasonable explanation as to why they have similar IQ's. Consequently, when we consider the question, "What is intelligence?" the answer essentially lies in the accumulation of data pertaining to the experience and environment of all former lifetimes. Present intelligence is the mental computer printout of those former lifetimes.

Now, let us turn to the question as to whether or not we can raise our IQ. Studies of deliberate efforts to raise IQ are almost non-existent, and are not contemplated in the foreseeable future. The reason for this lack of interest lies in the fact that most researchers have already accepted the acknowledged role of heredity in the determination of intelligence. They have concluded that after the age of seven, a person's IQ tends to stay about the same, and they subsequently imply that raising one's IQ is just about impossible. This last statement should have dictated that our adult activities remain childish or at the very best, show a strong resemblance to our infantile movements. This obviously is not the case.

Furthermore, the tendency to retain about the same IQ for the rest of our lives should more than convince us that we are subjected to some form of robotic consciousness[182] wherein the things we do may be beyond the control of human intelligence and free-will. Admittedly, our self-esteem and ego will never permit us to entertain such thoughts for fear that we may conclude that there is no thinking part to our brain. So, we ask, "Who are those biological researchers who would have us believe that the cerebral cortex is the thinking part of the brain, when it is believed that there is no thought process in the first place?"

The ongoing discussion of determinism versus free-will[183] is something that we lay people must begin to understand and reconcile. People all around the globe are interested in change. People want to take matters into their own hands, rather than leave it to government or other social agencies. In a sense, this is an encouraging sign because it means that the man in the street has come to realize that much of his future health depends more and more upon his personal decisions and lifestyle choices.

The 1980s found a civilization ready to surrender everything for the pursuit of personal pleasures. Restriction was uncommon and just about forgotten. The rape of Earth's natural resources continued unabated without considering the price that must ultimately be paid for this extravagance. Values diminished, especially for human life. The 1990s have, on the other hand, brought us to the realization that it is futile to attempt to solve society's problems by only pointing to manifestations, such as wages, housing, and crime.

The main error in the symptomatic attitude that steers the helm of government, business, and medicine today is its failure to ask the question, "Why?" Instead of directing energy toward curing the entity as a whole, attention is directed toward an endless patchwork process in which the outward symptoms are treated while the metaphysical cause is neglected. The Inner Light lies within the confines of the metaphysical level.

The kabbalist has long understood that to solve any problem we must see it in the context of a greater whole. To comprehend the microcosm, it is necessary to consider the macrocosm. Yet, this information and its rewards have been withheld from the general public.

We return to our original question, "Can we raise our IQ?" The answer is clearly in the affirmative. Our Inner Light (mind) consists of, and remains bound up with, the *tikkun* process. However, whenever we have achieved a correction— where in a prior lifetime negativity prevailed and this time around we exercised proper Restriction[184]—we remove a veil or *klipa* surrounding our Inner Light. At any given moment in our present lifetime, when the opportunity for Restriction presents itself and we succeed with our *tikkun*, the removal of one *klipa*[185] reveals a higher level of consciousness or intelligence of our particular Inner Light.

Intelligence, as defined by the kabbalist, is our limited aggregate or universal capacity to think, act, and deal effectively in our corporeal realm. The intelligence of each person is considered on a limited basis inasmuch as the ability and capacity to receive is governed by the individual soul-consciousness level necessary for that person's *tikkun* process.

Consequently, each person must, of necessity, enter this world with an Inner Light providing the minimum energy-intelligence to deal properly with his or her *tikkun* process printout. There is never a time when the Inner Light intelligence cannot cope with an impending *tikkun*, namely correcting a situation that one failed to do in a prior lifetime. And with each successful encounter, where this time around Restriction became the overpowering activity, the individual is rewarded with an elevated state of consciousness, over and above the original Inner Light intelligence he or she was born with.

If, for example, in a prior lifetime or lifetimes the individual committed a particular crime on a certain day of his 20th year, the opportunity for correction will again present itself on that same day during his present twentieth year. In exercising Restriction this time, the *klipa* that veiled the Inner Light consciousness is removed. Thus, by virtue of this Restriction, a higher level of intelligence is achieved.

Thus, the exceptions noted in many studies that indicate the possibility of raising one's IQ are not really *exceptions at all.* They are examples of those people who have demonstrated Restriction in their lives and thereby revealed a higher level of intelligence. These higher levels exist potentially from the moment of birth, within all of humankind. We all possess a gold mine from which we can extract everything we need to live joyously and abundantly!

227

Most of us are sound asleep because we are not aware of this wellspring of infinite intelligence. The Kabbalah can provide a link with the hidden power of our subconscious mental computer. We do not have to acquire this power. We already possess it. Within the depths of our soul-consciousness lies an infinite supply of all that is necessary for our physical and mental well-being. Throughout the wisdom of the subconscious mind, we find the solutions for every problem. When control of our thoughts is an accomplished fact, our conscious state, controlled by body-consciousness, can no longer interfere with its fragmented, distorted influence.

The ongoing struggle between the unconscious and conscious presents the possibilities for change. To remove confusion and limitation, we must remove the cause, and the cause is our conscious mind. In other words, the way we think is the way we act. The shepherd must lead the flock, and not the flock their leaders. *The conscious mind must be subject to the authority of the subconscious soul-consciousness.*

When the conscious mind is overcome with worry and anxiety, the negative emotions that result must be reversed. Soul-consciousness knows nothing of chaos, disorder, misfortune, and their related negative effects. Soul-consciousness is the channel and expression for the all-embracing Lightforce. Soul-consciousness is the seed and is more powerful than body-consciousness. But due to the teachings of Bread of Shame, soul-consciousness must restrict its awesome power when body-consciousness elects to lead us down the path of worry and anxiety.

For the present, it must certainly appear that universally, body-consciousness is firmly in the driver's seat. Considering the destructive behavior of humankind, body-consciousness enjoys an undisputed, unopposed march into the very fiber of human existence. The kabbalistic teachings support the human endeavor to halt the never-ending human carnage and suffering by connecting with the awesome omnipotent Lightforce that governs all things, above all, the body-consciousness.

As we look around, we notice that the vast majority of mankind lives in a world of robotic-no-control-consciousness. Little wonder that misfortune and degenerative diseases are the trademarks of our civilization. While our body-consciousness is in touch with limitation and fragmentation, our subconscious soul-consciousness, an extension of the Lightforce and infinity, speaks to us in the form of intuitions, creative ideas and the urge to share.

Writers, artists, and musicians tune in with their subconscious mental powers and thus become inspired. Mark Twain admitted on many occasions that he never really worked a day in his life. All his great writings, wit, and humor were the result of his tapping the inexhaustible reservoir of his subconscious mind.

The interaction between the body and our two states of consciousness—soul and body—determines the state of our health. Whenever the soul-consciousness prevails and dominates body-consciousness, we are then in tune with the intrinsic principle of harmony and continuity. The average child born into this world is usually perfectly healthy with all its organs functioning in excellent condition. We should be capable of maintaining this normal state and remain healthy, strong individuals throughout the course of our lives.

To be ill is to be considered abnormal. What this means is that we are going against the stream of soul-consciousness when we act and think negatively. We have become drawn into the abyss of our body-consciousness. There is a basic law of life, which goes as follows: If we entertain thoughts and activities that are not in accordance with the principle of "love thy neighbor," then these thoughts and activities—body-consciousness— eventually bring about disease and misfortune.

If we are in harmony with our soul-consciousness, then we increase the input and distribution of the vital Life-forces of our subconscious throughout our system. The study of Kabbalah provides this vital connection. It helps to eliminate our thoughts of fear, jealousy, hatred, anxiety, as well as our negative activities, which break down and destroy our vital immune system, causing nerves and glands to be impaired, and resulting in an overall degeneration of our vital organs.

We are the captains of our soul-consciousness and masters of our fate. Then why and when do we become vulnerable to the misfortunes and illness brought about by a pervasive body-consciousness? Our *tikkun* process establishes the arena for a free-will encounter with destiny. At any given moment, we may be called upon to decide whether to exercise Restriction or respond with a negative action.

If restraint is our course of action, then we have, for that particular *tikkun* process, removed ourselves from the vulnerability of becoming influenced by our body-consciousness. We have shut the door to and eliminated the consequences of body-consciousness domination. At that point, we have created another set of blueprints whereby soul-consciousness therapy is *now* permitted to restore to normal the functions of the body.

Today, more than ever, we are concerned with the process of health. Health costs have risen to a point where less and less people can afford health care. The fear of illness and concern with healing our bodies is on the minds of everyone. Yet we rarely ask, "What is it that really heals? Where does this healing power come from?" No physician has ever healed a patient. Doctors proceed to remove the blockages or barriers in the patient. A surgeon who removes the physical impediment has permitted the healing power to function normally.

The healing process is the Life-force within our soul-consciousness. Upon removing the barriers that inhibit and obstruct the flow of the healing process of the Life-force, healing then becomes a natural process.

These blocks and barriers originate with the body-consciousness, and the opportune time for its activity is when failure at the *tikkun* process has taken place. The negative influence of the body-consciousness then invades our body by the opening that we, ourselves, created with our negative thought and activity. Nevertheless, the Kabbalah can support us in restoring the powers of our soul-consciousness.

The degree of our healing depends on our ability to restore our individual Inner Light to its fullest revealment and potential. Each step along the journey of life is permeated with our *tikkun* procedure. Opportunities abound galore for increasing our IQ, which is nothing more than an increase in our energy-intelligence.

Throughout the ages, men of all nations believed there resided a healing power that could restore health. Healing of the sick was said to be possessed by the High Priest and other holy men. The "laying on of hands" has long been known by the kabbalists as a channel for the transference of energy.[186]

We can become our own healers. The Lightforce will flow through us, if only we permit it to do so. At the same time, we have the free choice of interfering with the normal rhythm of bodily functions, such as the heart, lungs, liver and other vital organs. Therefore, we must carefully watch our behavior and actions, and even more so, our dealings with our fellow man.

If, indeed, the Kabbalah provides the advantages and possibilities for the enhanced Inner Light mentioned previously, then let us now turn our attention to the more comprehensive energy-intelligence known as Encircling Light.

Encircling Light begins where Inner Light consciousness ends and extends beyond the Inner Light consciousness of humankind. Encircling Light is the all-pervading consciousness of the cosmos where information of past, present, and future meet as one unified whole.

Physicists are the first to inform us that we see but a fraction of what goes on around us. Even with the most powerful telescopes, we can see but a *tiny* portion of the universe, and conversely even the strongest electron microscope reveals only an *infinitesimal* fraction of the entire spectrum of atomic activity, and absolutely *nothing* of the subatomic realm. An apple would have to be expanded to the size of the Earth to see one of its atoms with the naked eye. Furthermore, beneath that atomic world is yet another world, the ratio of which is even greater than that of the atom with the physical world.

So when kabbalists tell us that the vast majority of what goes on in this universe is beyond the realm of finite understanding, they know well of what they speak. But then again, "Why is it necessary or even prudent to consider that which we can never see?" The kabbalist seeks to understand the source of all things. To accept the observable world as the totality of existence is to cheat ourselves out of the vast majority of life's possibilities.

Simple observation should tell us that the final manifestation of any event has nothing to do with the truth. The term "pull the wool over your eyes" as well as other expressions from the common vernacular, such as "snow job" and "smoke screen," imply a covering over of the truth. To the kabbalist's way of thinking, our entire phase of existence is covered over and concealed by negativity (*klipot*), and hence is deemed illusionary.

The *Zohar* and Rabbi Isaac Luria (the Ari) gave us a system by which to penetrate the crust of illusion that surrounds this world and find the infinite reality, the Encircling Light, within. No longer need we accept, at face value, the lies that pose as truth. Instead of being enslaved by deception, we can become the masters of our fate.

In this observable world, this tiny fraction of the spectrum of existence, we would be hard pressed to find anything with even the faintest resemblance to the truth. Indeed, the kabbalist will tell you that looking for truth in this world of illusion is like trying to find a subatomic particle in a haystack.

Our five senses are notoriously bad judges of the world around us. We all have, no doubt, been in a situation in which a sound is heard and every person in the room believes that sound came from a different place. The sense of taste and the closely related sense of smell can easily be fooled by chemical scents and additives. Nor is our sense of touch any better at gauging actuality. Any number of college pranks involving a blindfold, an ice cube, and the suggestion of fire can prove this. Taste, touch, smell, sight, hearing—all of our senses play tricks on us. Why, then, do we place so much faith in them? Where do we turn to find the truth?

There is a great deal in the *Zohar* that captures the imagination. Can we really simplify the universe's apparent complexity into a single thought-intelligence? Yes, states the *Zohar*.[187] The two fundamental and seemingly opposing forces that manifest in innumerable ways, including the seeming attraction and repulsion of the poles of a magnet, are not really distinct forces. They are rather different manifestations of the same underlying interaction that exists in the realm of the Encircling Light.

The cosmic glue, the single unifying energy-intelligence that governs all interactions in the cosmos is known by the code name *Masach DeChirik*, the Central Column.[188] Its name is Restriction, the amazing cure for all of the ills of both the Celestial and Terrestrial Realms.[189]

Can it really be that simple? The idea that we can reduce the staggering visible complexity of the universe to its essential simplicity by the power of our thought-intelligence is, to say the least, an exciting possibility. The words of the *Zohar*, "as above so below," go a long way toward describing a universe where all manifestations, physical and metaphysical, are tied together in a web of interconnected relationships, each apart from and yet part of the all-embracing unity.

The spiritual essence of the Hebrew *Alef Bet* emanates from the highest realms of energy-intelligence. The *Alef Bet* is not only permeated with the Life-force of the Creator, but also it is sealed with the impression of His Signature, which is Truth.[190]

The *Zohar* says:

> *Happy is the portion of he who calls the King and knows how to call Him properly. If he calls yet knows not upon whom he called, the Holy One, blessed be He, keeps away from him, as written, "The Lord is near to all those who call upon Him."*[191] *To whom is He near? It says again, "to all who call upon Him in truth" (Ibid.). Is there anyone who calls upon Him falsely? Rabbi Aba said, "Yes; it is he who calls yet knows not upon whom he calls."*

Whence do we know that? From the words, "to all who call upon Him in truth." What is "in truth"? It is the seal of the King's ring, which is overall perfection.[192]

This brings us to the question: "Why is knowing the Life-force essential to connection?" Furthermore, the *Zohar* stresses that the negative consequence of not knowing is that the Life-force *actually withdraws* from the individual. When one attempts to connect with any kind of power-force, a basic metaphysical understanding of energy and vessel must be acquired. Understanding of fundamental metaphysics is an integral part of communication, without which *true* connection may never materialize. Scripture in Genesis says:

And Adam knew Eve, his wife; and she conceived, and bore Cain.[193]

According to the *Zohar*,[194] the use of the word "knew" indicates sexual intercourse. But the idea raises many questions. Why does the verse use the word "knew" to describe the art of sexual connection? What is the real meaning of this coded message? Why this expression, when similar Hebrew words would have explained the passage more explicitly?

The *Zohar* explains that what we discover in this profound verse is a true connection with metaphysical forces dependent upon the knowledge that is derived from the establishment of proper channels. Knowledge is an integral part of this communication system because without it, any system is ineffective. Thus when Adam "knew" Eve, he established a method of clear communication through which metaphysical energy could flow unimpeded.

Therefore, the coded word "truth" implies knowing just what it is. And the *Zohar* makes this crystal clear when it states that it is the Life-force of the Creator that is essential towards the realization of any objective. Truth, then, is the embodiment of the Life-force.

There is another facet, states the *Zohar*, to the basic concept known as "truth." The *Zohar* recognized an essential flow when describing the Life-force as the fundamental ingredient of truth. The acknowledgment of and the desire to act in the Name of the Life-force or in the Name of the Lord has brought forth a history of murder and suffering. There is no more singular concept about which so much strife, friction, disagreement, and warfare has emerged. All religions act and represent themselves as the defenders of the Lord.

Then how, questions the *Zohar*, can one determine whether one's perspective is bound up and connected with truth, whether "One is calling with the signet of the King."? The seal with the Impression of His Signature, which is truth, is the code for the Central Column, which is the balance of everything. So it is stated:

> *This is the meaning of, "You will show truth [Central Column] to Jacob, loyal love [Right Column] to Abraham."*[195] *This is why it is written, "to all who call upon Him in truth."*[196] *And whoever knows not to call upon Him with the*

> *quality of the Central Column, but tends to the Left*
> *Column or the Right Column, the Holy One,*
> *blessed be He, draws away from him.*[197] *Happy is*
> *the portion of whoever entered [wisdom] and came*
> *out whole, to know the ways of the Holy One,*
> *blessed be He.*[198]

The central idea that emerges from the *Zohar* is man's ability to connect with the whole or quantum of the universe by exercising the principle of the Central Column. The quantum of the universe, where yesterday, today, and tomorrow represent themselves as one single unified whole, is what the kabbalists refer to as the Encircling Light. The superconscious of the universe is what Encircling Light consciousness is all about.

Superconscious Encircling Light takes off where Inner Light consciousness ends. The all-pervading consciousness of the cosmos—where information of past, present, and future meet as one unified whole—extends beyond the Inner Light consciousness of humankind. It is precisely this consciousness —Encircling Light—that we find most wanting in our lives.

The element of surprise, which usually creates much upheaval in our lives, can be avoided when connection with the Encircling Light exists within our consciousness. The irrational components that have not been taken into account by our consciousness are, nevertheless, elements upon which our well-being depends. Once we begin to recognize the limitless feature of superconscious Encircling Light as an essential feature for the development of a wholesome life, then surprises no longer constitute an element of confusion or bewilderment. We are suddenly thrust into a thought consciousness whereby we begin to appreciate the idea of wonder and astonishment.

In achieving a higher frame of awareness, we realize the restricted vision of our five senses. A quantum perception is beyond the grasp of our finite consciousness. Activity originating in faraway places most certainly affects the activities of an individual, notwithstanding the distance between them. Consequently, the future of any plan must remain at some point in jeopardy of uncertainty and with it, the likelihood that if anything can go wrong, it will.

To achieve a relationship with superconscious Encircling Light will require many techniques in the kabbalistic process. The idea of mastery of our destiny involves a mastery of our universe and universal activity. By any stretch of the imagination, our present thinking on this matter is severely limited.

For the most part, we have not found any viable method to insure our physical and mental well-being. There are too many factors involved preventing definitive results from becoming materialized. We act and pray, mostly pray, that our decisions will bring forth the desired results. Then again, who can say for sure that the results we seek are those best suited for our needs? It certainly does become confusing.

How much time and effort is consumed by the unknowable? Our despair and frustration is incalculable. Connection with the superconscious Encircling Light adds a new dimension, namely that of precise determinism. But everything depends on whether or not we apply the restriction-sharing energy-intelligence of human activity or instead choose to indulge in the Desire to Receive for Oneself Alone. Selfish activity creates a program of uncertainty in which even the most seemingly flawless plans become subject to the decision of the quantum universe.

When we act with the energy-intelligence of Restriction, we can then access the superconscious Encircling Light as the program for our daily existence. When this occurs, along with the kabbalistic techniques, everything becomes improved, even beyond our best laid plans. The superconscious removes all the rough edges and replaces any doubts or uncertainties.

What emerges from the *Zohar* is the potential dual activity that is present in all energy-intelligence. The *Zohar* drives home the point that if our negative human activity prevails, then a severance takes place between the continuity phase of reality and our human plans and hopes. The kabbalistic techniques plus the positive energy-intelligence of Restriction are our links to certainty, to the realm of Encircling Light. The implications and benefits of this quantum superconscious are profound for ourselves and for the world.

Chapter Seven

IMMUNE–VULNERABLE

Chapter Seven

IMMUNE–VULNERABLE

THERE HAS BEEN A GREAT DEAL WRITTEN IN RECENT YEARS about our immune system and why some of us are more vulnerable than others to its breakdown. Here again, the ills and misfortunes of bad luck remain mysterious when we explore the reasons why some, and not others, become exposed to its fury as bad luck creates a path of destruction in our lives.

The victims of drug and alcohol addiction are described as weak-minded people who fall prey to influences over that which they have no control. Yet kabbalistically speaking, people who succumb to drugs are looking for a free adventure into spirituality without the use of Restriction. Hence, as the need for more and more stimulant drugs increases, the real need is less and less fulfilled.

Medical experts and psychiatrists want to explain states of illness or addiction as they *appear*, without exploring the question of why some people are affected and others not. It is unfortunate that these experts do not examine the role of cosmic influence and reincarnation. If we do nothing more than chart these patients' activities in daily life, together with negative cosmic danger zones, we will find some exciting differences between the way in which the so-called lucky ones become and remain immune while the so-called unlucky ones become vulnerable.

There are three special differences to be investigated. Firstly, the lunar-solar months in which people are born. Secondly, when and if negative behavior and activity took place during these cosmic danger zones. Thirdly, and most importantly, the security shield against attack from the evil negative energy-intelligence, the force of the Dark Lord.

Before examining the mystery of those phenomena, let us first explore the kabbalistic viewpoint regarding the idea of a metaphysical security shield, the likes of which seems to exist only in science-fiction novels. A good starting point for our investigation is, as always, the *Zohar* itself.

> *When Jacob departed from Laban, all the holy legions surrounded him so that he was not left by himself.*[199] *Rabbi Chizkiyah asks, "If this is so, why is it written, 'And Jacob remained alone?'"*[200] *Rabbi Yehuda replies, "Because he exposed himself deliberately to danger, and therefore the guardian angels [the security shield] deserted him. It was to this that he [Jacob] alluded when he said: 'I am unworthy of all the mercies and of all the truth, which Thou hast shown Thy servant.'"*[201] *...Rabbi Elazar said, "The sages have stated that on that night and at that hour, the power of Esau [negative energy-intelligence], was in ascendant and therefore Jacob was left alone [meaning vulnerable]."*[202]

Assuming that we become familiar with those negative points of time, how then do we avoid falling into the trap that has been laid for us? Let us, again, turn to the *Zoharic* text that can and does provide all of humankind with the necessary tools to deal with those unseen negative energy-intelligence forces.

> *Rabbi Shimon opened a discourse on the verse: "Better is he that is lightly esteemed, and has a servant, than he that plays the man of rank, and lacks bread."*[203] *This verse, he said, speaks of the Dark Lord, the evil prompter, who lays plots and unceasingly brings up accusations against man. He puffs up a man's heart, encouraging him to arrogance and conceit, to carry his head high, until the Dark Lord obtains dominion over him. Better, therefore, is one who is "lightly esteemed" and who does not follow the evil lord but remains humble in heart and spirit.*
>
> *The evil prompter is bowed down before such a man, as it cannot control him. On the contrary, as it is written: "But you may rule over him."*[204] *"He who is lightly esteemed" is exemplified in Jacob, who humbled himself before Esau [Dark Lord] so that the latter [Esau] should in time become his servant.*[205]

If we are to become masters of our destiny, then let us learn the cardinal rule: An *effort* is required on our part. We have become a society seeking instant relief simply by paying for it. This approach leads us down a road towards disaster. There are no easy or shortened methods to achieving permanency in our well-being. To avoid the work and the responsibilities that come with this mastery, we are drawn to capsulized measures that only proclaim their temporary relief.

All spiritual techniques, including Kabbalistic Meditation, serve as a secondary application towards improving our physical and mental well-being. The primary and initial step that must be taken is a resolve to develop a positive attitude towards our fellow man and environment. With the cosmos filled with negative energy-intelligence created by man's adverse activity, we find it very difficult to resist and operate against the stream of negativity. Meditative techniques will go a long way in providing assistance in overcoming these obstacles. But this mastery requires the knowledge and resolve that, for us to change the quality of our lives, we must change the nature of our behavior.

Misfortune, in whatever manner it appears, is the direct result of our negative attitudes of present or former lifetimes. That we can change the direction of our misfortune is what the study of Kabbalah is all about. However, Kabbalah distinguishes itself insofar as the prerequisites for achieving an improvement in our well-being is concerned. We must put forth the effort! The Kabbalah knows quite well the timetable during which the Dark Lord has been given dominion in our universe, and provides the necessary security shield that can and will protect us from his devastating onslaught.

Embracing negative attitudes is essentially connecting with and embracing the Dark Lord. The creation of an affinity with this dark environment prevents any progress in the removal of misfortune, whether it appears in the form of ill health, family problems, financial instability or other calamities.

This now brings us to an understanding of vulnerability. There are two conditions that must be met to guarantee mastery of our lives and destiny. Firstly, our attitude towards our fellow man must be one of humility, so strongly suggested by the *Zohar* just quoted: Jacob understood all too well, that for him to overcome the essence of the Dark Lord, he needed to "humble himself before Esau."

Secondly, we must don a security shield when negative energy-intelligences prevail within the cosmos, despite the fact that we may boast of wholesome, positive attitudes. Referring to the previously mentioned *Zohar*, Jacob had a serious encounter with the Dark Lord. Although Jacob was decidedly the victor, he nevertheless suffered a severe blow to the thigh.[206] He had exposed himself, albeit deliberately, to danger.

We learn from the preceding *Zohar* that even Jacob, the chariot and embodiment of the Central Column,[207] became *exposed* to the onslaught of the Dark Lord's legions. The lack of a security shield demonstrated the vulnerability of even so powerful an individual as Jacob.

What seems to emerge from the *Zoharic* text is at least an understanding of the enigmatic question as to why one person is vulnerable and another is not. Take, for example, the cosmic effect and influence of lunar Cancer. When scientific researchers develop the theory that "happy people do not have cancer," they touch upon an accurate result of their investigation. However, as is customary in the biomedical paradigm, the physical, manifested state is usually taken as the cause rather than the result.

Granted that happy people avoid the scourge of cancer, the question that we must ask of ourselves is, "Why are some people happy and others not?" In essence, we must explore and search out the essential cause. When the whys and wherefores to all and any answers are resolved, we may then assume that we have come upon a primary cause.

There are essentially two basic developments that contribute to mankind's vulnerability. Both are channeled through the cosmos or celestial bodies. Before we investigate the agents responsible for our becoming vulnerable, let me make it clear once again that mankind potentially is in a position to convert his vulnerability to immunity. The problem of vulnerability is not, from a kabbalistic perspective, a question of who will fall into the category of being lucky or unlucky.

We become exposed to the misfortunes of vulnerability when in past lifetimes we had failed to overcome our Desire to Receive for Oneself Alone.[208] Exposure to traumas and misfortunes that accompany vulnerability take place in this lifetime, exactly within the same time frames that we experienced this same exposure in prior incarnations. If in a previous lifetime, we failed in our *tikkun* process to restrict the urge to steal, then, in this lifetime, we will experience another opportunity of free choice as to whether or not we will overcome the urge to steal and succeed by Restriction.

If we fail, then we have created fragmentation, limitation, and "empty space," the non-revealment of the Life-force. This gap represents the energy-intelligence of "vulnerability." Our body-consciousness, constantly on the lookout for tapping the negative energy-intelligence of the Dark Lord, finds its crack in our defense lines and attracts the enemy Dark Lord to be drawn into us. The invasion by the Dark Lord has breached our natural, birth-given defenses and immunity systems, and has thus given rise to the Dark Lord's negative energy forces.

Once negative energy forces have established a beachhead within our system, these forces begin to spread at highly increased rates of attack. They continue to fan out throughout the natural immune system, establishing pockets of resistance to any attack by the immune system.

Vulnerability, therefore, is where the future has been shadowed by the past. Vulnerable people are endangered by a genetic "time-bomb" passed from a former lifetime to the present. When a particular group is at risk because of disorders that may start with a single missing or defective gene, these diseases, from a medical standpoint, are traced to flaws in our DNA as the primary cause. Hence, the concept of heredity is adopted, where symptoms are carried through generation after generation by descendants. This lethal legacy must be traced beyond the basic blueprint for life. We must ask ourselves, "Why was I picked to be in such a family?" and "What does it have to do with my *tikkun* process?" It is not by chance that we were put in this position. From outward appearances groups of people are banded together by family, geographic or ethnic ties, so doctors and researchers present a solid case for the heredity or common denominator theory. Nevertheless, the point established by kabbalistic teachings is that group relationships are not primary causes, rather these people have been banded together because of prior lifetime *tikkun* ties. A physical demonstration or manifestation is *never* considered the primary cause. It may point to the cause but cannot be considered as primary.

With this in mind, present flaws in DNA or general misfortunes are the direct result of failure at the *tikkun* process. For every action, there is a similar and opposite reaction. The hurting of others results in damage to the perpetrator. Consequently, when the opportunity (misfortune) presents itself again, the cassette dictating the oncoming of a reaction can and will be altered by a demonstration of Restriction. This plays the major role in the body's immune or disease fighting system by inducing the production of antibodies.

It is no strange coincidence that the immune's fighting system is comprised of *antibodies, opposed* to the body. And what is the body-consciousness if not the Desire to Receive for Oneself Alone? Producing these natural antibodies, the soldier cells that defend against illness, is the task assigned to the consciousness of Restriction.

Consequently, positive moods of Restriction[209] can have a subtle, yet telling, effect on our immune system. Does this mean an attitude underlined by Restriction can be valuable in fighting diseases? Although the *tikkun* process predisposes us to vulnerability, the restrictive factor allows us to overcome the predetermined life cassette.

The time, place, and circumstances for incarnated vulnerability is manifested by an orchestration of the cosmos. The channel by which this exposure to the Dark Lord becomes manifested is the cosmos. Therefore, a well-made and read natal chart can be an invaluable tool by which the astrologer can pinpoint these specific gaps and opportunities for the *turnaround* of our *tikkun* designation.

But in the absence of a cosmic timetable, to play it safe under any circumstances when impending misfortune comes our way, we must treat this incident as an opportunity. Restrictive energy-intelligence can stave off the effects of misfortune and illness. An ounce of prevention is worth a pound of cure.

Another cause of the breakdown or damage to the immune system is exposure to various negative cosmic danger zones. The responsibility to avoid these time frames rests with all of us. Ignorance of these devastating cosmic events will not prevent their invasion of a healthy body. *Zoharic* teachings reveal that cosmic predisposition can play a role in infinite illnesses. Kabbalists zero in on these cosmic markers with sufficient information to map the location and time frames.

The use of such markers to predict disease is enhanced by data extracted from the world's largest and most comprehensive collection of metaphysical knowledge, the *Book of Formation*, the *Zohar,* and the *Writings of Rabbi Isaac Luria.* This may sound unbelievable but some day, we will be able to predict, at birth, everyone's vulnerability to disease and scientifically validate these findings. But at present, we can only inform the kabbalistically-oriented people of their inherent risk factors. The idea that our bodies and vital organs are bound up and connected with cosmic time zones is clearly stated in the *Zohar*.[210]

> *And in the first part, the Faithful Shepherd started by saying, "Woe to those people whose hearts are closed and whose eyes are unseeing, who do not know the parts of their own body and according to what they are arranged.*

Knowledge is the connection[211] to the control over our bodies and over illness when it strikes. It therefore behooves us to make every *effort* to relate and understand the teachings of the *Zohar* in *Parashat Pinchas*. At the same time, we must become aware of those mistaken individuals who continue to lead us astray by shunning the teachings of the *Zohar*, claiming that the *Zohar* is for the select few alone. The *Zohar* says:

> *And Lilith [do not pronounce this name] is called spleen, and she goes to play with the children, later killing them, and makes of them anger and tears, to bewail them. The spleen goes to the right of the liver, which is Samael [do not pronounce this name], who is the Angel of Death. This, that is the liver, was created on the second day of the Work of Creation, while the other, that is the spleen, was created on the fourth day of the Work of Creation. And for this reason it is not a good omen to commence something*

*on Mondays or on Wednesdays. Liver is death for
adults; spleen is death for children.*[212]

What seems to emerge from the foregoing *Zohar* is the
intimate relationship between cosmic zones and our body. Yet
little attention is paid to these zones, particularly their threat
to the well-being of our mental and physical health. The *Book
of Formation*, the authorship of which is attributed to
Abraham the Patriarch, abounds with the body-cosmos
connection. I acknowledge that this subject is alarming. It
suggests that everything about our medical procedures needs
to be carefully examined and possibly modified. But we
cannot ignore the fact that what happens at birth may pave
the way for vulnerability in later life.

When a 15-year-old Maryland boy walked into his parents'
garage and hanged himself, the entire community was
stunned. When a 13-year-old boy in Mexico City set himself
afire, a confused neighborhood asked why. These were
students who had never run afoul of the law. These
youngsters seemed to be coping well with the usual pains of
adolescence. To shocked friends and family, their deaths were
inexplicable and tragically premature. To psychiatrists and
psychologists studying the phenomenon of teenage suicide,
they were just another statistic in the growing epidemic of
teenage suicides.

There are several scientists who suspect that a tendency toward
suicide, as well as vulnerability to drug addiction, may be
linked to traumatic birth experiences. When considering the
overall suicide rate in the United States, which has tripled in

the last 25 years and has increased most dramatically among young people, we must make every effort to determine why.

According to the National Center for Health Statistics, one American teenager attempts suicide every 78 seconds. The statistics are chilling, and the idea of attributing this epidemic to the stress of modern life does not sit well with many a researcher. After all, the stress factor of our society affects all and not just some. Obviously statistics do not tell us *why* young people are taking their lives.

It is, therefore, encouraging to read that some researchers are convinced that traumas suffered at birth appear to be imprinted within the unconscious. These traumas might be responsible, they say, for a compulsive urge to repeat the trauma during adolescence. In the 1920s, psychoanalyst Otto Rank proposed a similar theory relating birth experience to neuroses. This idea was quickly denounced by his peers who commented that birth traumas make about as much sense as astrology.

However, from a *Zoharic* perspective, birth traumas do relate to astrology. In fact, birth traumas are the direct result of astrological influences that suggest the physical expressions of the birth traumas themselves. We are constantly reminded of the kabbalistic principle that physical manifestations or expressions may never be used to determine primary causes. Unfortunately, most theories regarding the circumstances underlying adolescent suicides, as well as other mysteries of human existence, continue to focus on psychological and social phenomena.

Consequently, sociologists researching the problems implicate everything from the intense pressures of contemporary society to heavy-metal music to violent films. They fail to take notice of the metaphysical, pre-physical thought realm. After all, do we indulge in any physical activity without prior thought? If this is a most important concept, namely that the metaphysical precedes the physical and the metaphysical is constantly moving the whole of the cosmos, then why do researchers fall into the trap of claiming that physicality is the primary cause?

Jerusalem has been noted as the Holy City. Is it because the Holy Temple was located there? The kabbalistic retort to this kind of reasoning is: "Why was the Holy Temple located in Jerusalem in the first place?" The physical expression of the Temple cannot determine underlying causes, inasmuch as we are left with the question: "What brought about the Temple site in the first place?"

The *Zohar* documents the link between the Temple and the City of Jerusalem. The initial cause for the Temple's location is related to the energy center of the universe abiding in the Land of Israel. The *Zohar* tells us that the Temple and the Ark within the city of Jerusalem were collectors and conductors of cosmic energy-intelligence. When a circuit of energy flowed, the universe and all of its infinite galaxies were in harmony and violence did not exist.

The Temple of Jerusalem reflected something entirely different from the temples of other ancient peoples. What happened in Jerusalem affected everything on Earth and in the cosmos. Jerusalem was considered to be the nucleus around which all galaxies revolve. As ridiculous as this may sound, conditioned as we are by the narrow contemporary view of science, Jerusalem did not, nor does it now, represent some religious ideology. Physical structures merely symbolize a metaphysical thought energy-intelligence and portray its internal realm. The body reflects the internal soul-consciousness of humankind. The body is a secondary force. Our soul energy-intelligence is what motivates the body-consciousness.

The sooner we come to this realization, the better the possibilities for reaching meaningful and lasting solutions to our problems. The present approach to problem solving has done little to provide for the objectives necessary to enhance the well-being of our society. Superficial reasoning in determining cause and effect is a convenience of the five senses. This reasoning permits us to sit back comfortably, thinking we have worked things through. Little do we realize that, as Murphy's Law implies, "If anything can go wrong, it will." Gaps or cracks in our reasoning are easily filled or replaced by the negative world of the Dark Lord.

Every avenue must be pursued in determining primary causes, despite the sometimes frustrating efforts to get to the "bottom of it all." Complacency and feelings of shallow security are, like the ostrich, false guarantees that our problems will fade away. The entire spectrum of quantum reality must be researched before coming to any conclusions. The answer to the final "why" provides the ultimate cause, and that final definitive cause is our gateway to problem-solving.

Let us, therefore, return to the serious problem of adolescent suicides and the scientific approach to its implications. The scientific method is clearly in question inasmuch as the search for some answers always falls within the biomedical paradigm of a superficial, physical external character.

An intensive research program conducted in 1987 by Dr. Bertil Jacobson, at Stockholm's Karolinska Institute has come up with data indicating that events surrounding an individual's birth may influence his later decision to kill himself. Jacobson's results were even more specific than he had anticipated. In one U.S. study, he found that suicide was more closely associated with birth trauma than with any other of the 11 risk factors for which he tested, such as socioeconomic variables, parental alcoholism, and broken homes.

In a study conducted in Sweden, he found a correlation between the *kind* of birth trauma and the *mode* of suicide. For example, he found that those who asphyxiated themselves, whether by hanging, gas poisoning or strangulation, were four times as likely to have suffered oxygen deficiency at birth. Twenty percent of those who chose to end their lives by mechanical means, such as guns or knives, experienced mechanical trauma, such as breech presentation or forceps delivery at birth.

His study dropped an unexpected bombshell. He confirmed that future addicts were born in hospitals where doctors freely chose to administer barbiturates and other drugs to women in labor. As for women who chose drug-free childbirth, their offspring would not nor should be vulnerable to drug abuse.

But this admittedly is not the case. Furthermore, science is faced with another dilemma. How can an episode of physical trauma or brief period of exposure to a drug produce self-destructive behavior later in adolescence?

Most of us cannot remember anything that took place in those early years after birth, let alone traumatic births. There have been some suggestions that for some inexplicable reason, traumatic experiences suffered at birth appear to be *imprinted* in our brain. This imprinting is responsible for a compulsive urge to repeat the trauma during adolescence. Therefore, some people choose to go through life with a brain made insensitive by drugs or an urge to choke themselves with a rope imitating their first day of life when an umbilical cord wound around their throat threatened to cut off their oxygen supply.

The real problem as seen through the eyes of the kabbalist is: "What prompted the mother to have the particular traumatic birth in the first place?" "Why do some expectant mothers choose drug-free childbirth and others do not?" Vulnerability or immunity. These two words encompass the basic causes that we have presented in this chapter.

Because I do not want to leave you hanging without some answers to this pressing problem, let us turn to the source, the *Zohar*. The insight and pursuit of an original cause is what Kabbalah is all about. There is no way that we can successfully deal with situations of epidemic proportions unless we can trace them back to the original cause.

However, before acquiring the ultimate understanding of life's experiences, there is just one other point worth mentioning. There is a growing awareness among psychologists today that although the baby's brain is not fully developed at birth, the activities of the mind, separate and apart from the brain, begin to explore and pay close attention to each new experience. All of this happens despite no clear explanation as to how the mind can operate independently from the brain. There also is the essential, most important, question regarding the *relationship* between traumatic births and this urge to repeat the trauma during adolescence.

As I just mentioned, a baby's brain is not fully developed at birth. Yet the achievements at the infancy stage are to be considered almost miraculous. Babies cry for food, crawl, and walk, a feat that is never duplicated as we grow older. We almost stare in amazement at the journey each child takes from helpless infancy to self-expression. The random babbling of babies turn into words. From baby steps to gymnastics, we witness miracles that have thus far eluded scientific understanding. Our elusive mind does not and cannot satisfy the questions that have just been raised. Let us now turn to the *Zohar*.

> There is a commandment to assign the death sentence, which is four in nature: by the sword, by strangulation, by stoning, and by burning. To whom does the Bible address itself? To Samael [the Dark Lord].[213]

In a few words, the *Zohar* provides us with a penetrating insight into the devastating traumas of uncommon deaths. These four unnatural means by which people die encompass all other forms of traumatic deaths of mankind. They all have one thing in common. If an individual has succumbed to the energy-intelligence force of the Dark Lord by virtue of not restricting a particular Desire to Receive for Oneself Alone, and this particular failure is of such a serious nature as to warrant the consequences of one of these four modes of death, then when the soul returns in the next incarnation, a traumatic birth will signal the kind of death the person underwent in a previous lifetime.

I cannot repeat often enough the awesome power of information provided for mankind by the Bible when decoded by the Kabbalah. Anything and everything has an essential, underlying reason for why and how things happen. However, our five limiting senses, plus conventional mind programming, prevent us from groping deeper into the matters that affect our daily existence.

The argument against predictability in nature is a basic tenet of quantum theory. Heisenberg's Uncertainty Principle establishes an inherent, inextricable *interminableness* in the web of the micro-world. Subatomic events have no methodically definable cause. At the root of this dilemma, from the kabbalistic perspective, lies human free-will. The subatomic spectrum, and this includes man's metaphysical energy-intelligence, is impervious to physical laws. This explains how the energy-intelligence of man can and does wreak havoc in the cosmic flow.

Consequently, we have been drawn to the erroneous belief that what we cannot relate to does not exist. If the cause cannot be defined or determined, then the assumption is that some events are to be treated as occurring without any apparent cause. Science does not possess the scientific tools with which to regress to that time that preceded birth. However, the Kabbalah *does* provide the missing information, thus providing an in-depth, all-pervasive picture concerning the original cause and its effect.

Therefore, a traumatic birth that involves the incident of near strangulation by its umbilical cord symbolizes this soul's nature of death in a prior lifetime and possibly the cause of death in its present incarnation. Furthermore, the lesson that must be learned and drawn from this experience is the particular circumstances and events that brought about the soul's failure in the past lifetime.

Before concluding this chapter on vulnerability, I feel compelled to raise the subject of natural disasters as they are perceived through the eyes of the kabbalist. This subject has, at best, been left alone, lest we invite the wrath of the religionist who declares that, "The Lord in His mysterious ways knows and has His reasons for raining destruction upon various parts of our globe."

This idea or conviction does not sit well with most of us. Does the Lord really play dice with Earth's inhabitants? Additionally, how and what decided where particular fault lines were to be located? The San Francisco earthquake of October 1989, was an event that seemed part biblical and part Hollywood. And with the alarming prospect of the Big One, the October 1989 shattering earthquake may one day be remembered as a dress rehearsal for the Big One. The future quake, according to seismologists, has a 50 percent chance of striking the Bay Area within the next 30 years. If and when it does, California will suffer catastrophic losses of life and property.

Earthquakes may be acts of the Lord but man's ability to deter, delay or even prevent one from taking place is not even considered. We are simply vulnerable for we are not in a position to prevent them or even predict them with any degree of precision. Seismic forecasts are expressed as odds because geologists are privy to so few of Earth's secrets.

Although the rest of the country is not immune to earthquakes, California gets more than its share because it is riddled with more fault lines than a raisin. Will there be any warning? Curiously enough, we are not the masters of our planet, a fact that we ignore at our peril.

Let us, therefore, turn to the *Zohar*, which opens to the scientist of tomorrow, as well as the layman of today, a new understanding of the laws of nature. The *Zohar* repeatedly states, "…*that which is below is above and that which is above is below.*"[214] Reinforced with the theory of relativity, everything is really connected and interdependently related. The revelation of the *Zohar*, with careful scrutiny of the relationship between the individual and his environment, allows us to learn exactly what exists in the realm of the galaxies as well as in the realm of undiscovered and concealed objects.

Rabbi Yitzchak once drew near a mountain and there beheld a man sleeping under a tree. He sat down. Suddenly and without warning the earth began to quake violently and became full of fissures. The tree was uprooted and fell to the ground. The earth was rising and falling.

And the man beneath it woke and wept, and lamented with mourning and sound of sorrow. For at this moment a great Supernal minister is now being appointed in Heaven, who will cause terrible misfortune to the world. The quaking of the earth is meant as a portent and warning to you. At this time, Rabbi Yitzchak felt a trembling and said: Verily it is written, "For three things the earth quakes, for the servant when he reigns... For an odious woman when she is married, and a handmaid who is heir to her mistress."[215] That is, rulership has changed hands and the kingdom will now pass on to the more evil one.[216]

What seems to emerge from the *Zohar* is that earthquakes are the result of the fall of kings, queens, and political leaders replaced by representatives of the evil side. The Dark Lord and the community of evil consist of the life-energy force of Desire to Receive for Oneself Alone. The *Zohar* cautions: "Man, by virtue of his evil deeds [succumbing to the Desire to Receive for Oneself Alone], accrues power to the death star fleet."[217]

As a result of man's negative activity, the power of the Dark Lord is greatly enhanced and the result becomes manifest through terrestrial disasters. The quantum link between man's behavior and his environment determines the behavior of his surroundings. Thus it becomes evident that certain areas of our globe, more ripe and vulnerable to inexplicable disasters, are dependent on the behavior (positive or negative) of its residents.

Fault lines are not the primary factors of earthquakes but rather the inevitable result of negative human behavior. Were the activities of mankind reversed to one of Restriction and Sharing, the fault lines could no longer dominate the California coastline. Las Vegas would no longer live in fear of becoming the new western surf of the United States. At this time, I see no other hope or possibility for the California coastline residents than to begin their daily scanning of the *Zohar*.

In conclusion, vulnerability cannot be defined as "unlucky." The Quantum Theory forced us to see the universe as a single web of thoughts and interactions strung among the various parts of a unified whole. Therefore, it is now time to take a hard look at the negative atmosphere generated by many movie and television productions. The external and internal aspects of the individual are woven into an inseparable net of mutual interconnectedness of all entities and events. These aspects can neither be ignored nor treated as separate entities.

The most important characteristic of the kabbalistic world view is the awareness of the unity and mutual inter-connectedness of all entities and events. The Dark Lord's force and vision of the universe is one of fragmentation and destruction, and his activity alone would insure a world enslaved by the death star fleet. However, the conscientious scanning of the *Zohar* can develop the necessary protection of the security shield, while all around us things totter, crumple, and fall.

The trend towards greater individual control of our life and destiny has recently become more pronounced among some medical professionals. There is even some agreement that so-called primitive medicine often achieves remarkable cures with a few herbs administered in *dogmatic* ritual. The ceremony involving these herbs was designed to arouse the individual consciousness and strengthen the patient's belief in himself or in a higher power. These procedures are based on the knowledge that, in the final analysis, it is the patient himself who brings about the healing.

In the March 1989 issue of *The Lancet*, a prestigious British medical journal, a study on the survival of breast cancer patients was presented. Their findings showed that 80 percent of those patients diagnosed as having a fighting spirit had a 10 year survival rate, whereas only 20 percent of those tagged as feeling helpless had a 10 year survival rate.

The report underscores the dilemma that has arisen in modern medicine. We do not foster the patient's natural healing abilities. Unfortunately, there is no required course for the medical student on how to deal with people or encourage their natural healing abilities. The medical profession is preoccupied with the fear of death in which the foe is disease and death means failure.

We neglect our innate ability to heal ourselves. We are expected to give ourselves up to current medical practice. Yet at Johns Hopkins Hospital in Baltimore, Maryland, breast cancer patients who were non-conformists and had a poor relationship with their physicians were longer-term survivors. What this meant was that the patient who was assertive and shared responsibility in her recovery was considered a bad patient but achieved better results. The response of the wise patient to the doctor who claimed that he was the captain of the ship was, "I'm not sure I want to board the doctor's ship."

A change is coming. Surgeons no longer freely assault and invade our bodies at will. Oncologists will not poison to kill the disease but will rather consider the patients' abilities to heal themselves. The transition is coming more swiftly than any scientist currently imagines.

Another example of man being relegated to a position of inferiority and mindlessness is the introduction of the modern Golem, the computer.[218] We have become so dependent on this animated creature that we have all but forgotten how to add or subtract. The computer does our programming, to the extent of even programming our lives. Today, we find ourselves so enslaved by this massive machine that we have ceased to think or function as higher human beings. Will we remain at the mercy of these aliens?

The *Zohar*, therefore, is a refreshing relief for those lost in the maze of computer printouts or in the anxiety of pacing the barren cold corridors of hospitals. While change is slow in coming, it is nevertheless coming. The basic tenets of the

Kabbalah have always claimed that man bears the sole responsibility for what happens to him and his environment. Man has been given the awesome power to influence his entire planet and cosmos.

This idea is strikingly revealed by the *Zohar* when considering the Biblical narration of the Great Deluge, the Flood:[219]

> *"And the Lord saw the Earth and behold it was corrupt; for all mankind had corrupted their way upon the Earth." How can the Earth become corrupt? Is she governed by the rules of reward and punishment? The answer is that since man governs and influences the Earth, his behavior and actions, if they are evil, imbue their evil spirit within the land. When mankind commits sin upon sin, openly and flagrantly, then the Earth becomes brazen-faced and behaves accordingly. Hence, this is the connection here: "The Lord saw that the Earth was corrupted." Why? "Because all flesh had corrupted their way upon the Earth."* [220]

What seems to emerge from the foregoing *Zohar* is the primary cause behind all "so-called" natural disasters. Earthquakes, floods, tornadoes, and hurricanes are all the "natural" result of man's evil inclination and behavior. If we have pinpointed man as the underlying cause for these disasters, then it follows that man can alter or even prevent these catastrophes from ever happening in the first place.

This idea also provides us with some kind of rationale as to who will become the survivors and who will not make it through. After all, there have always been the so-called survivors who were "lucky." Why those people and not the others?

The notion that we are in a position to thwart natural disasters is clearly stated in another section of the *Zohar* that discusses the matter of earthquakes:

> *It behooves man to consider and learn the machinations of the Lightforce. For the mass of mankind does not know or reflect upon that which keeps the world or themselves together. For when the Lightforce created the world, He made the Heavens of fire (Left Column) and water (Right Column) mingled together but not compact or in harmony with each other.[221] Only later, were they brought together as one unified whole by the power of the Central Column. From this point onwards the Lightforce planted these supports for the terrestrial world including Earth. When the force of Central Column is removed or not activated, they all quiver and shake and the world trembles. As it is written, "Who shakes the Earth out of her place and the pillars thereof tremble." When the force of Central Column is drawn to the Earth, then the world is supported by the Three Columns and thus the world remains in a balanced state and does not quake. "When the channel of the Torah (the letters of the Alef Bet) is drawn into the world's scenario, everything remains in a supported state."[222]*

In conclusion, the *Zohar* makes it perfectly clear that what has always been considered "acts of the Lord," is a corruption of the creative process. Natural disasters are our own doing. We have the power to prevent and stop these unnatural phenomena from taking place. And if the world around us is all that evil, then as Noah before us, we can take matters into our own hands for the safety of ourselves and our families.[223]

Noah was taught how to construct a security shield that would isolate him and his family from the coming devastation brought on by the flood. The Ark was a symbol and manifestation of the all-powerful Lightforce by which the hail and brimstone of the flood waters were kept at bay. While everything around Noah perished, he remained alive. This remains a morbid lesson for our generation, which has become exposed to so many "natural" as well as man-made disasters.

We do not have to hope and pray each day that we will be counted among the "lucky ones" or the survivors. We can and must do something about it. There is nothing else left for us to do but scan the *Zohar* and be certain that our activities and behavior include the forces of positivity and the Central Column.

Chapter Eight

TIME TRAVEL

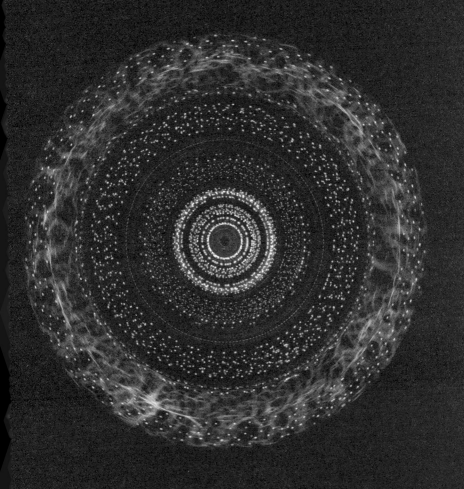

TIME TRAVEL

ONLY RECENTLY HAVE SOME MAINSTREAM PHYSICISTS politely admitted that time travel, once the exclusive domain of science fiction, is at least possible in theory. However, the design of a time machine is not under consideration for the present. Scientists have enough problems trying to work out the kinks in time travel theories. The key point, at this time, is that physicists have found nothing in the laws of physics that would *prohibit* time travel.

There is the possibility that someday someone might somehow do what science fiction characters do now: Speed through hyperspace to a distant galactic outpost traveling faster than the speed of light. Scientists tell us that if we could travel at the speed of light, time would actually proceed into the past. An astronaut traveling at the speed of light could possibly zoom into deep space and return before he left.

Time slows down for a moving object, as measured by an observer considered to be stationary. According to this theory, if one member of a pair of identical twins travels at light-speed to a distant star and back, he will return to find that *his* clock has been running slower and that his *homebody* brother has aged more than he has.

This time-expansion effect has been confirmed by experiment. Atomic clocks taken on long jet rides have been found to lag at five billionths of a second behind clocks that remained on Earth. The established laws of science must be reconsidered when a scientist talks about an elastic time that can be extended or reduced, stretched or shrunk, as well as places where time no longer exists or where subatomic particles travel back in time.

Although such forms of time may *appear* to be unacceptable to most of us, nevertheless they open the doors to even more and stronger phenomena, and challenge our most rigid laws and principles of logic. Mathematical formulae *suggest* two-way time travel, yet for the most part, the paradoxical consequences keep it within the framework of science fiction. Why?

The most common flaw presented against the idea of two-way time travel has become known as the Grandfather Paradox. A time traveler would encounter it if he returned to the past just in time to prevent the meeting of his grandparents. This, in effect, means he, himself, never was born. But had he never been born, he could not have prevented the marriage of his grandparents or have been present to go back in time in the first place. In other words, if time travel is possible, how do we avoid violating causality?

When someone goes back in time and thereby is in a position to affect his own past, things could get sticky. Suppose I arrive at a spaceship terminal at 4:45 PM and board at 5:00 PM to fly back in time through the time tunnel. At 4:45 PM, the spaceship returns from its flight in time to see me arriving at the terminal. Can I decide at that time not to take the flight? If I do, how could I see myself returning from it?

Then we must consider also how far back in time will the time machine take the traveler. For the present, scientists can think only in terms of the time dilation effect slowing down time. Scientists maintain it cannot make time march backward. At its very best, time becomes frozen. There is no chance that we are going back to visit the dinosaurs.

Things do get a bit sticky, however, when someone or something that has gone back in time is in a position to affect someone's past or even his own past. How can something go back to the past and affect its future in such a way as to prevent its backward journey through time?

The most popular proposal made by science fiction writers is that somehow we are prevented from doing anything in the past that would affect our future in a way that cannot be reconciled with time travel. This proposal simply contradicts the very idea of time travel to the past. In other words, they are proposing that you "cannot have your cake and eat it too." But this, in essence, is just another way of saying we really have not come up with any answers to the initial problem.

There is, of course, the kabbalistic idea of parallel universes. This is the concept in which a person can affect the future but ends up in a parallel universe where the future is different from the one he or she came from.[224]

Physicists are not impressed with the idea of parallel universes despite mathematical formulae suggesting two-way time travel.

The essential problem for the physicist-mathematician is that this phenomenon challenges our most rigid laws and principles of logic. If we could travel faster than the speed of light, then we must consider the possibility of stumbling upon the Fountain of Youth. We reverse the steady pattern of time passing into the past and thus eliminate the reality that the cosmos is slowly disintegrating or that people must grow old. The conclusion about the inevitability concerning entropy leads mankind, as well as scientists, to conclude that we *cannot* accept the concept that nothing passes away or that there is no past time.

This discrepancy is resolved if we accept the reality of two parallel universes. While scientists and philosophers argue about the nature of time and question its very reality, most of us still continue to think of time as the duration of everyday processes. But Earthly-time has been well documented as elastic time or time dilation. The closer we get to the speed of light, the slower our watch will run. If we could reach the speed of light, time would stop altogether.

In light of these strange effects, one minute for some will now be two minutes for others who have not achieved a closeness or integration with "faster than the speed of light." Einstein's theory of relativity takes us beyond our experience, which is the basis of common sense and logic. For how logical is it to believe that if we were capable of traveling at the speed of light, that we would then return prior to our departure?

If one could travel faster than light, time would actually run backward. We would take a trip and return the day *before* we left. This idea snaps back at our brain and logic like a rubber band. Our normal reasoning seems to melt away before our very eyes. A theory of such far-reaching implications, running counter to our daily experiences, could never have made it into the mainstream of scientific thought had it not been often verified in many laboratories.

Relativity once and for all shattered the commonly accepted viewpoint of an absolute time. It opened the doors to another reality of flexible time that is completely dependent on the state of *motion* of the observer. Kabbalah goes one step further and declares that physical motion is just the tip of the iceberg when it comes to the reality of flexible time.

The state of our mind is yet another form of time travel where time goes haywire. Words like "sooner," "later," "now," and "simultaneous" are relative expressions. What is here and now for one, is there and then for another. From the perspective of a man who is late for an important appointment, time is rushing by at a breakneck pace, while from the point of view of another man who is early for the same appointment, the same time may be dragging on interminably. How time is perceived depends on the perspective from which it is observed.

If the man who is late for the appointment were to be suddenly teleported to his destination, time would immediately be transformed from a restraining tyrant to a benign servant, worthy of accolades instead of diatribes. Time would have brought all of the parties together at the same time, in the same place, for the same meeting.

While it is true that most of us cannot stem the tide of "time marches on," we can change our perception of that statement. By so doing, we can significantly alter the course of our lives. Imagine time as a river that runs from the far distant past into the far distant future. Imagine that the flow of the river is controlled by your wants and needs, moods and emotions. When your thoughts are clear, so are the waters of time. When you are agitated, the waters are also agitated. When you are in a hurry (as was the man who was late for his appointment), the banks of the river are narrow and the waters are white-capped rapids. When you are at rest, the waters run cool and calm. When you are afraid, the waters are dark and ominous. When you are at peace, the waters are mirror-glass smooth.

What, then, is the essence of time? Is it a friend or an enemy? Is it our servant, a mere convenience by which we measure our lives, or is it a tyrant who rules over us with an iron hand? Do we use it or does it use us? There is no single answer. In the final analysis, time is what *you* make it.

With the foregoing revolutionary idea of time behind us, we may now find little difficulty in accepting the kabbalistic version of time and time travel. If we could travel faster than light, time would actually run backward. We could go on a trip and return the day before we left. We could also be traveling back to the days of our *youth*. In effect, we have found the Fountain of Youth, namely, our *own* particular fountain of restoration.

The myth surrounding the search for the Fountain of Youth has always left us with the idea that it appears somewhere outside of ourselves. However, if time ran in reverse—and physicists find no reason why it should not—dead people would come alive, trees would "un-grow," and broken glass reassemble. In fact, physicists have yet to agree on a theory that explains why time does go forward.

Let us now begin our journey back in time in our quest for a better future. Going back in time via our time-travel machine is, in reality, traveling towards the future as well. Because our journey begins at this very moment, even though we are traveling back in time, the trip still follows our present moment in time. Consequently, we are moving towards time in the future or alternatively phrased, we are moving *back* but in a sense *towards* the future.

Yet from a kabbalistic perspective, and I might add from a scientific point of view, everything that has ever *happened* in the universe, everything that ever *will happen*, and everything that *is happening* in the present, has been unalterably determined from the first moment of time. The future may be uncertain to our minds but it is already arranged in every minute detail.

And it is precisely this conclusion that brought about the differences that exist between Kabbalah and the yet unverified suppositions that science so blindly follows. Scientists have resolved that therefore *no human decisions* or actions can change the fate of a single atom, for we too, are part of the physical universe. However free we may feel, everything that we do is, according to the scientist, completely determined. All of existence is thus encapsulated, frozen into a single moment. Past and future have no real significance. Nothing actually happens.

The problem that science must acknowledge is that if the arrow of time can point either way, then trying to un-break an egg, grow younger or make a river flow uphill should be a perfectly acceptable sequence of real events. These events *do not* take place because the physical processes that occur in our world seem to be irreversible. You simply cannot make things go backwards. Furthermore, if the underlying laws that govern the activity of each atom are reversible, what is the origin of its irreversibility?

However, the kabbalistic world view maintains that human decisions can change the fate of an atom.[225] The physical world as we know and see it conforms to the uncertainty principle of illusion. This universe exists as the "gathering place" for our illusionary universe. It is here that time, space, and motion make their presence felt. Within the realm of this "uncertainty" universe exists the fragmentation of time where events seem to be irreversible and becoming younger simply does not happen.

The necessity for our illusionary universe stems from the original *Tzimtzum* (Big Bang or first Restriction), our desire to remove Bread of Shame, our ambition to experience a participation in the creative process. In the Endless or reality level, the future does not exist as a separate entity. Growing younger indicates a creative process of reversibility. The fallacy behind the idea of becoming young again is the fact that we have never aged to begin with. There is never a future period of aging, nor for that matter, was there ever a past that saw us becoming older. There is nothing more to reality than the present, which contains the supreme level of wholeness and certainty with the illusionary by-products of uncertainty, chaos, and deterioration.

We are told by medical authorities that aging and deterioration are natural processes that are irreversible. However, their conclusion is not based on any scientific data other than what we observe as physical reality. This opinion contradicts scientific laws and principles that maintain the existence of reversibility. The kabbalist states that the apparent aging process is yet another example of the illusionary realm in which we live.

We now return to our original question, "How do we become younger?" Or stated another way, "How do we maintain the perfect state of our existence? How do we capture the developmental stages from infancy to adulthood without paying the observable price of aging?"

As we mentioned previously, the problem of returning in time requires a time-tunnel. For the present, science is still exploring the possibilities but as yet to no avail.

Firstly, where within the *Zohar* is this concept of returning back to the past mentioned? And secondly, what, if any, time capsule does the *Zohar* make available to achieve this phenomenon? Before proceeding to tackle these heretofore unanswerable questions, we must give more thought to the machinery that generates the Life-force that makes life possible. We must begin by raising the questions whose answers scientists have been struggling with for centuries. Those answers still remain in the dark.

The reason that we should address ourselves to the mysteries of our universe is that the answers are affected by our ability to ask the questions. After all, what came first, the question or answer? Since questions precede answers, then it follows logically that within the question already lies the answer. The seed always includes that which follows and evolves from itself.

We have come so far in the pursuit of the sciences, yet with all our wonderful discoveries, we are no closer to controlling our destiny. What has the universe got to do with my everyday lifestyle? How is the pursuit to understand the cosmos related to the enhancement of my mental and physical well-being? Now that we have been successful in popularizing science, what import does it have on my daily life? Although certain types of surgery are now more effective because of the advent of laser technology, how does that affect me?

It is precisely at this point in time that the Kabbalah is arousing interest. The Kabbalah raises questions about everything. It zeros in on the ideas that explain *everything* from before the Big Bang to mankind's power to control the universe and consequently to control his own destiny.[226]

For this reason alone, the *Zohar* has commanded such respect for almost 2000 years. Its awesome power has led some people to be fearful of delving into its secrets. They were afraid of being harmed by its mystical holiness because they were not spiritually pure to handle its awesome knowledge. This fear, according to many kabbalists, was well founded up until the middle of the 16th century. From that period onwards, which kabbalists consider to be the beginning of the Age of Aquarius, the limitations and prohibitions surrounding kabbalistic study were completely removed.[227]

In our society, it is still the practice of teachers and parents to shy away from most of the questions raised by the *Zohar*. Many find themselves uncomfortable with the issues raised by the Kabbalah. In anticipation, they fear the demands and responsibilities that may be thrust upon them when they discover that, indeed, there is the possibility of obtaining quality control in their lives. At the same time, we have also succumbed, although incorrectly, to the computer which so visibly exposes the limitations of our human abilities and understanding.

The *Zohar* is the breath of fresh air necessary to revive us from our deep and long slumber. Einstein once stated to an interviewer, "All I want to know are the Lord's *thoughts* on how He created the world, for the rest are only details." The *Zohar* embraces both the thought and reasoning behind everything, including precise details. The ultimate theory of the universe presented in the *Zohar* is consistent with everyday observations.

The *Zohar* brings to an end a long chapter in the history of humanity's intellectual struggle to understand the universe. But most importantly, the *Zohar* also revolutionizes our consciousness, thus enhancing the quality of our daily lives.

Let us now turn to the *Zohar* for some insights into time-travel and the time-machine that provides our journey back to the past, the machine that can readily take us from a disordered, disoriented present to an orderly arrangement of the past.

Time is moving at an ever-increasing rate of speed. In an amazingly short span of time, we have developed from the day of the horse and buggy to an era of space travel. To what can we attribute this phenomenal change?

According to the *Zohar*, "*all the celestial treasures and hidden mysteries that were not revealed to succeeding generations will be revealed in the Age of Aquarius.*"[228] It is noted that the New Age will provide us with a comprehension, not only of our familiar universe but also of that which lies beyond the range of observation in the realm of the metaphysical, the non-space domain.

Today, we are witnessing the beginning of a new age of revelation. Today, more than at any other time in history, the Lightforce is demanding to be revealed.

> *As the Lightforce separated those on Mount Sinai from all limitations of our mundane, physically expressed world, so shall the Lightforce separate them at the Final Redemption (the conclusion of the Age of Aquarius).*[229]

The Revelation on Mount Sinai is interpreted by the *Zohar* to mean a connection between the raw, naked energy of the

Lightforce and mankind. Hence, the use of the word "revelation," which means the Lightforce being revealed without the usual protective elements that conceal and provide, and thus diminish and dilute, the awesome power of the Lightforce.

With the removal of the illusionary, corporeal realm of existence, the impediments that prevent or slow down our movement, cease to exist. At the time of Revelation, the idea of speed of light no longer has any reference. Movement is instantaneous once we consciously decide where we want to be. Space, as we know it, has no place once our illusionary, physical realm disappears. Time, as we know it, flies out the window, and past, future, and present become elevated to one of unity with the Lightforce.

> *The generation of the Exodus even saw all the future generations of mankind up to the days of King Messiah.*[230]

This startling declaration by the *Zohar* reveals that events can operate not only from the past to the future but also from the future to the past. The Lightforce, therefore, is our time machine, our entrance to the higher worlds.

Only the Lightforce is capable of removing the illusion of corporeal reality, revealing a cosmic model that is, was, and will always be, timeless and full of certainty. This was the phenomenon of Revelation. When Israel fell under the influence of the Golden Calf,[231] its connection with the Lightforce came to an end. The Israelites could no longer harness the awesome power of the Lightforce, and eventually they perished in the wilderness.

Revelation was and is an opportunity to connect with the proper tools and channels for achieving the altered state of consciousness that enabled Moses to connect with the Lightforce. The Lightforce was revealed. There was no turning back. The awesome power of the Lightforce was too much for mankind to handle at that period in time when Revelation took place. The Lightforce became a manifested entity and state of consciousness in the physical world of reality.[232]

This time around, however, following and during the beginning of the Age of Aquarius, mankind will once again be given the opportunity to connect with the Lightforce. Once more, we will have the chance to experience the awesome opportunity of connecting with the Lightforce. This connection will result in our ability to travel back in time or stated another way, reach and achieve an altered state of consciousness where the past and future are here now, where our youth is again upon us, where we will benefit from the Fountain of Youth, where death has been terminated as a physical part of our landscape.

Time-travel or traveling at speeds of light was not an uncommon experience for kabbalists in the past.[233] The fundamental problem facing scientists today lies in the inadequate propulsion to achieve or exceed travel at the speed of light. The *Zohar* does not consider the problems connected with *approaching* the speed of light as the singular obstacle in achieving human travel at speeds faster than light travel. The solution lies, not in effectively producing propulsion that will approach the speed of light beyond the light barrier but rather in simply *removing the barrier itself.*

Eerie as this may sound, it is all in perfect harmony with the new age ideas concerning the laws of space and time.[234] Amazingly enough, the *Zohar* presents a plan that will provide man with the capacity of turning the entire solar system, including planet Earth, into a human backyard. This idea is no less incredible than the effect of the airplane that converted the once formidable world oceans into little more than swimming holes.

The answer, as so clearly and simply stated in the *Zohar*, is that the *removal of the light barrier* depends completely on the *removal of the mankind barrier*, represented by mankind's hatred and intolerance for one another. This feat of overcoming the obstacles of space travel and the light barrier was clearly demonstrated by both the Prophet Elijah and Pinchas, who were actually one and the same person.[235] Both knew and understood where to look in the hazy forest of light barriers and thereby knew how to dissect the anatomy of interstellar flight.

The removal of physical barriers depends completely on our ability to remove our metaphysical barriers of intolerance and hatred. However, until such time when the whole of mankind will recognize the necessity of removing this obstacle, those individuals, whose inherent character and lifestyle relate to the concept of "love your neighbor as yourself," must, should, and can avail themselves of the opportunity to connect to the awesome power of the Lightforce revealed on Mount Sinai.

Let us, therefore, turn to the *Zoharic* revelation of time travel and the time-machine necessary to remove the light barrier. But before we begin, let me explain what the removal of the light barrier actually means. This might be compared to the digging of a tunnel where the excavation is accomplished by a mechanism that disintegrates and vaporizes the earth and stone, thereby permitting the tunnel to be completed in the little time it takes this spectacular digger to travel from one end of the tunnel to the other.

This same principle applies to the *Zoharic* time-machine. By vaporizing the light barrier, which is the substance derived from the Desire to Receive for Oneself Alone, the concept of speed, movement, and space disappears. This is the exact situation the astronaut encounters when he is in outer space. The barrier that we experience here on Earth—some refer to it as friction—is more dense and concentrated than the barrier in outer space. Consequently, the astronaut can travel up to speeds of 24,000 miles per hour. This is still, of course, not as fast as the speed of light because the light barrier, although less than on Earth, exists in outer space as well.

However with the *Zoharic* time-machine, the essence and level of the light barrier (friction or the Desire to Receive for Oneself Alone) is broken down and vaporized. Consistent with the space-time continuum, the illusion of space and time become nonexistent. The separation of one person living in the United States and another person living in China is measured by the space or distance between them or the time it takes to travel between the two points. Time and space, however, are illusionary measurements of guidelines. Therefore, once our vaporizing time-machine acts upon the light barrier, distance and time as we know it does not exist. We are instantly in two places at the same moment. We have succeeded in being in the past, as we are in the present. The past for us is, in fact, already the present, indistinguishable and undifferentiated.

> *And when the Scroll is to be brought to the altar for reading on the Shabbat, it is incumbent upon all present to prepare themselves in awe, fear, trembling, and in sweat. And to meditate in their hearts as if they were standing now on Mount Sinai to receive the Torah Scroll.*[236]

The passage in the *Zohar* confirms and shows time as symmetrical. As the equations of the physical sciences express time as symmetrical (meaning the equations work as well in one temporal direction as in the other), the *Zohar* finds no difficulty in stating the conditions required to move in reverse time. As we just mentioned, reversed time means returning to Revelation at Mount Sinai where the Lightforce became manifestly expressed.

The prerequisites stipulated in the *Zohar* are quite similar to the fiction (soon to become non-fiction) movies that portray individuals traveling back in time in a time-machine trembling, sweating, and fearful. Severe physical stresses cannot be avoided during flights into space and back to Earth.

Heavy vibrations and the forces of rapid acceleration start the journey off with intense strains. These are the same identical conditions mentioned in the *Zohar*.

Concerning the speed of light factor, if, as the *Zohar* suggests, we are to condition ourselves for the flight back to Mount Sinai when Revelation took place, then in a matter of seconds we have traveled approximately 3300 years, faster than the speed of light. Once we have reached speeds faster than the speed of light, science agrees that we can move in reverse time. We have indeed achieved the phenomenon of traveling faster than the speed of light. Having accomplished this feat, we now become the beneficiaries of all the benefits that reverse-time has in store for us.

The explorer in H.G. Wells' novel, *The Time Machine*, built a device that moved through time but remained in the same place. From a scientific point of view, no known temporal phenomenon can do that. In the scientist's comprehension of time-reversal, there is a change of position as well as change in a point in time. Time travel for the scientist means transcending that limited sense of temporal journeying, either moving backward or forward in time. In either case, an object would be transported out of the light frame into a region that is neither here and now, nor past and future. Time travel in a practical sense is still elusive, and time still seems trapped within the barrier of light.

Wells' novel, *The Time Machine*, was essentially fiction in nature. But then again, so were many other tales of fiction which eventually caught the attention of fellow physicists. The general feeling among cosmologists is that unless some really advanced beings have already made a time machine, we are not going back to visit the dinosaurs. What is obviously more disturbing is that whoever builds the time machine could cause a lot of mischief. This machine could cast a cloud over all of physics.

But time machines, or time holes as some cosmologists call them, have always attracted the attention of, and tickled the fancy of, the science-fiction-hungry public. "Back to the future" has attracted too much attention outside the rarefied world of cosmologists and astrophysicists. The scientists find it very bothersome because if time travel is possible, then how do you avoid violating causality? Or, to put it more dramatically, "What would happen if someone went back in time and killed his own grandmother?"

The issue is here to stay and will never again fade away. Physicists are not uncomfortable with the concept of going back in time. Richard P. Feynman, a famous Nobel Laureate and Professor of Theoretical Physics at the California Institute of Technology in the United States, once showed that positrons, the antimatter counterparts of electrons, could be regarded as electrons that are moving backward in time. Kabbalists have no problems in dealing with this new phenomenon, which was previously explained by the idea of parallel universes.

There is a story told of Rabbi Isaac Luria (the Ari), that he wished to be in Jerusalem for the Sabbath and within minutes he arrived in Jerusalem from Safed, which is some

200 miles away. To read this story one might think that time in this context is an illusion, but we learn that when one goes above the gravity of Earth, time itself changes. There are many such stories in the *Zohar* of our sages traveling from one place to another, and they almost represent time as a variable.[237] The *Zohar* presents the concept of traveling through time as one of *remaining in the same place*. The physical body of those present in the synagogue for the Torah reading on *Shabbat* does *not* transcend into another realm or region. The body *remains* in the same place, which is the same idea presented by Wells. For all we know, he may have been familiar with the *Zohar*.

According to the *Zohar*, what is necessary is that the remainder of the human being, the 99 percent of us, our souls, the reality segment of all human beings, transcend and elevate to the realm of the true reality. How do we achieve this transformation? By transforming our Desire to Receive for Oneself Alone to the Desire to Share.[238]

Scientists view *time*, and what it means, the same way that we *consider* it. Objective time has gone. It has gone the way of all outdated scientific principles. Time has shown the inherent limitations of science. It has done this by taking us from the once important material world, the world that belongs to human experience, and bringing us as far as the signpost pointing to the true reality of existence. The signs are pointing to the world above the physical one, the world of consciousness, the world referred to in Genesis as the Tree of Life.[239]

The Tree of Life consciousness is the realm of pure awareness. The world of illusion, fragmentation, space, and time have no place within Tree of Life consciousness. The Tree of Knowledge consciousness is where we experience chaos, deterioration, and disorder. Once we experience the true reality of the Tree of Life, we have removed the illusion of time. We have at once accessed into, and have reached, every part of the universe where past, future, and present are one with the Lightforce. Hence, we remain in the same place.

When time ceases to be a parameter, we begin to enjoy the fruits of certainty and order. Within the Tree of Life realm, space does not exist, and it is similar to the illusion of time. The space-time continuum, developed by Einstein, merely substantiated that measuring of movement could no longer be reached by a time or space factor. Both time and space were no longer objective. Both time and space were truly illusionary, except that we are still led to believe that space-time are objective quantifying measurements.

The *Zohar*[240] strikingly demonstrates its position on time and the notion of speed of light. It maintains that space belongs to the illusionary realm of the Tree of Knowledge, where true reality is contained within the illusionary environment of life's experience:

> *And he took from the stones of the place, and put it under his head and lay down in that place... the land on which thou lies, to thee I will give it.*[241] *Said Rabbi Yitzchak, the verse teaches us that the entire land of Israel was condensed and shrunk to the size*

of Jacob's body. Hence the possibility that Jacob lay on the entire land, meaning the land of Israel.

Two great revolutions gave birth to the new physics: Einstein's theory of relativity and quantum theory. The first casualty resulting from these theories was the belief that time is universal and absolute. What Einstein demonstrated was that time is, in fact, elastic and can be stretched and shrunk by motion. The second casualty resulting from these theories was that space is also elastic. Few people would ever dream of the possibility that what is one foot today may be two feet tomorrow or that the same one foot of today may be one-half foot tomorrow. Yet not only does the theory of relativity demand that distances have no absolute and fixed dimension, it also suggests experiments to verify these discrepancies.

We all take for granted that we and all material things must be somewhere, someplace. When physicists began to explore the concept of location in the light of quantum physics, they were shocked to find that the very idea is meaningless. The basis of this dilemma is a fundamental rule, known as the uncertainty principle established by Heisenberg, the German contemporary physicist. He brought to the forefront the notion that things can be everywhere at once: There is no space, as the *Zohar* determined. An activity in space may very well have gone out the window.

Following on the heels of the *Zohar* declaration that space was indeed illusionary, I recall trying an experiment with a group of 150 people. We were going to walk to a place some 45 minutes away. The road we were to take had little, if any, traffic.

I suggested that we all lower our eyes and visualize the road passing under our feet rather than our feet trekking along and over the road.

Everyone immediately sensed a kind of *remaining in the same place* and the road *moving* under our feet. The 45-minute walk no longer consciously felt as if we had walked that long. Most of us experienced no motion whatsoever.

The problem that we must all face is the ill-advised programming that has become part of our lifestyle. Science snugly walked in the footsteps of trained and organized common sense. All its concepts were firmly rooted in the common sense world of daily experience. Time was time as we experienced it, and space was the culprit to overcome in getting anywhere we wanted to go. New revelations of physics were put on a firm foundation. And while new phenomena stretched science beyond the realm of direct human perception, these phenomena were still formulated as simple extensions of ideas and objects already familiar to us.

The new age of physics made its appearance around the early 1950s. The cozy notions of reality that had endured for centuries were blown away. Many unquestioned assumptions and cherished beliefs were shattered. Suddenly the world was revealed as a weird and uncertain place. Common sense became an unreliable guide. Physicists were forced to restate their conception of reality. More importantly, their notions and new ideas had no direct counterpart in human experience.

The old world view of a mechanistic and rational universe collapsed into oblivion, to be replaced by a metaphysical world of paradox and uncertainty. What was essentially wrong with the conclusions of the physicists was that the initial premise of Newtonian physics started with a flaw in that it failed to consider that we live in and usually experience a world of illusion, rather than, as Isaac Newton claimed, a universe subject to rigid laws of cause and effect. The concept of the illusionary nature of material existence was expressed by the *Zohar* and known by the kabbalists long before Newtonian physics made its entrance into the mainstream of human experience.

In fact, when the bizarre workings of the micro world replaced a mechanistic universe, space and time were cast into the realm of metaphysics. Space and time were to the human experience as air and blood are to human existence. Firstly, the problem now was, "How do we face up to a reality that seems to oppose our preconceived, rational notions of how we expect the world to act?" Secondly, the fundamental question for physicists and lay people now was, "How do we retrain our thinking after being told for centuries that there is no reality other than the physical one to which we usually relate?" Space and time never did, nor do they now, exert the influence our programmed brain was conditioned to believe.

Everything points toward a consciousness reality rather than a physical reality. But some phenomena seem so hard to believe or imagine that even eminent contemporary physicists refuse to accept them, either personally or professionally. After all, they too, are human and their own personal experiences of space and time are not consistent with the claims of modern physics. Einstein, himself, found these new ideas difficult to accept, even though his theory of general relativity opened the field to the new physics.

But then again, if we laypeople were fully aware of the dilemma facing the scientist, we would be up in arms demanding to know why so much money and research are devoted to a subject in which they themselves do not believe. How can these scientists continue their research in good conscience when common sense has collapsed in the face of the new physics? As it turns out, scientists are not quite as straightforward as they have led us to believe.

Scientists continue to lull us to sleep with the statement that Newton's laws are, of course, still good for most everyday phenomena. What is essentially wrong with this attitude is that it leaves us stuck in our same grim track record of strife, illness, misery, and constant global warfare.

Kabbalah maintains that we need not remain puppets in the hands of uncertainty principles. We have the possibility and the responsibility to regain control over our destiny. After thousands of years, we still find ourselves on a collision course with certainty, order, and happiness. We are now being programmed to believe that the scientists can achieve a better society for us to live in. For the present, they have dismally failed. After all our great achievements, we still cannot predict what tomorrow holds in store for us.

On a higher plane of existence, cautions the *Zohar*, sanity and order does reign.

Through the study of Kabbalah, the principles laid down by the *Zohar* are now accessible to all. We can indeed connect with the world of certainty and order, and leave behind the illusionary world of chaos and disorder. The *Zohar* leads us to forms of great simplicity and beauty that we have not yet encountered. Its teachings reveal a genuine feature of nature. Its wholeness and frightening clarity makes us all wonder why we had not thought about this before. We cannot help thinking that this must be true. Its creative imagination can produce a theory, so compelling in its elegance, that we become convinced of its truth before it is subjected to experimental testing.

The point that this book and the *Zohar* drive home is that we have an orderly universe around us. But before we can move ahead and have access into this universe, we must first rid ourselves of the belief that we are helpless human beings aboard a rudderless ship on a stormy sea. We can and must assure ourselves that we, and we alone, will master the future course of our life experiences. Life is not a game of chance. Chance is but an illusion.

And now to our final subject, the Time Machine. Here, too, we must turn for instructions to the master Kabbalist, Rabbi Shimon Bar Yochai.

If space and time are no longer well-defined in the quantum realm, known in kabbalistic terminology as the Tree of Life consciousness, it comes as no surprise that kabbalists such as Rabbi Shimon Bar Yochai, Rabbi Isaac Luria, and others overcame the artificial differences and fragmentation that beset the human experience.

This was not merely an exercise in "psyching the mind" to overrule the notions of space and time. For the kabbalist, the cosmic code of the Bible provided a method by which mankind could alter and change the life experiences of chaos, disorder, and uncertainty, known as the Tree of Knowledge universe. The Tree of Knowledge comprises all sorts of barriers, whether they be physical, emotional, psychological or imaginary. All of these barriers, referred to by some as friction, were placed in the Tree of Knowledge realm to prevent us from getting a "free lunch" or "free trip." We are required to remove these barriers, which kabbalists call the Bread of Shame concept.[242]

A central challenge to the barriers of space and time is the paradoxical nature of resistance.[243] The kabbalist, whose essential, internal character consists of a lifestyle that includes Restriction in every aspect of daily mundane life, finds the quantum leap or "tunneling" through the barriers of our physical world as easy as crossing a deserted road.

Taking the Time Machine requires the necessary preparations and precautions an astronaut must take before a flight into space. The *Zohar* states that these preparatory measures are the prohibitions of the Bible placed upon man that support him in his attempt to achieve a full complement of resistance or Restriction. When man is prepared, taking the Time Machine is a "piece of cake," with man enjoying all the pleasures and sensations of time travel.

Now you may ask, "Why do I take so long to make this point? Why have I kept you in such suspense to learn what is the

Time Machine to which the *Zohar* refers?" Firstly, if you have gotten this far, you have experienced the feeling of Restriction. Secondly, the Time Machine cannot and does not work for those of you looking for a free lunch. The sooner we recognize that our world is not a free universe, the better our chances of succeeding with our Time Machine.

> *Said Rabbi Shimon: When the Scroll is removed from the Ark for the purpose of public reading, the Gates of Heaven of Rachamim*[244] *are opened, initiating and awakening the Love of Above (the all Embracing Unified Whole) and then man should say the following: Berich Shemei ....*[245]

The prayer of *Berich Shemei* is the *Zoharic* Time Machine. Though we may never be able to experiment with the Time Machine directly, we can appeal to our inner consciousness, and ultimately to our physical body-consciousness, to determine how we feel and whether we experience the tunneling through the barrier of time and space.

The new physics continues to provide novel insights into the workings of the universe, where nature seems to play tricks on us. One of these is the barrier trick. Imagine throwing a stone at a window only to find that it penetrates the glass and appears on the far side, leaving the window intact. Yet this piece of trickery is precisely what electrons seem to do. In effect, they seem to tunnel through the insurmountable barrier.

The Kabbalah restores the mind to a central position in our universe. When we tunnel through space-time and travel at the speed of light towards Revelation and connect with it, we achieve the ultimate: The Tree of Life universe unfolds itself before our very eyes.

I contend that the thoughts presented here are revolutionary. Coming to grips with these seemingly outlandish notions does tax the imagination. History has shown us that the truth always turns out to be more wonderful than anything we can imagine. The universe appears to be full of activity. To the kabbalist, violent phenomena are simply expressions of human violence. Good and evil apply to the Tree of Knowledge universe. However, the kabbalistic journey can prepare us for entrance into the realm of the Tree of Life consciousness where chaos and disorder will be recognized for what they are—an illusion.

Kabbalah teaches us the way to remove ourselves from the spiritually impoverishing cycle of negativity, struggle, failure, and ultimate defeat.

Kabbalah leads us to a state of mind in which we are connected with the infinite continuum, where time, space and motion are unified, where past, present, and future are entwined, where everyone and everything are interconnected; where *then* is *now*, and *all is to the power of One.*

REFERENCES

INTRODUCTION

1. *Power of the Aleph Beth*, Vol. I, Berg, pp.29-38
2. *Kabbalah for the Layman*, Vol. III, Berg, pp.35-37
3. *Kabbalah for the Layman*, Vol. III, Berg, pp.35-44
4. *Zohar*, Lech Lecha 30:322
5. *Zohar*, Bo 13:203-204

CHAPTER 1 COSMIC ATTACK

6. *Zohar* II, p.l71b
7. Zohar I, p.53a
8. *The Modern Rise of Population*, Thomas McKeown, San Diego, Academic Press, 1976
9. Haggerty, 1979
10. Exodus 21:19
11. *An Entrance to the Tree of Life*, Rabbi Yehuda Ashlag, ed. Berg, pp.54-58
12. *Kabbalah for the Layman*, Vol. I, Berg, p.73
13. *Kabbalah for the Layman*, Vol. III, Berg, pp.l00-l01
14. *Kabbalah for the Layman*, Vol. III, Berg, pp.l00-101
15. *Kabbalah for the Layman*, Vol. I, Berg, p.93
16. Numbers 24:3
17. Numbers 24:2
18. *Zohar*, Noah 26:197-199
19. *Zohar*, Tetzaveh 2:31
20. *Time Zones*, Berg, Part III

21. *Wheels of a Soul*, Berg, pp.78-81
22. *Power of the Aleph Beth*, Vol. I, pp.65-89
23. *Kabbalah for the Layman*, Vol. II, Berg, p.31
24. *Power of the Aleph Beth*, Vol. II, Berg, pp.33-35
25. *Power of the Aleph Beth*, Vol. I, Berg, p.123-128
26. *Jerusalem Talmud*, Tractate Shabbat, p.14b
27. Numbers 24
28. *Writings of the Ari*, Rabbi Isaac Luria, Vol. 6, Gate of Introductions, Research Centre of Kabbalah, Section 2, p.31
29. Leviticus 12-13; Kabbalah Connection, Berg, p.142
30. Leviticus 16:21
31. Proverbs 22:9
32. *Zohar* Acharei Mot 20:124-128
33. *Zohar* I, p.165b
34. *Zohar* Beshalach 10:114 & 118-120
35. Leviticus 19:14
36. *Zohar* II, p.9b
37. *Power of the Aleph Beth*, Vol. I, Berg, pp.120-121
38. *Kabbalah for the Layman*, Vol. III, Berg, pp.148-149
39. *Power of the Aleph Beth*, Vol. I, Berg, pp.25-28
40. *Zohar* III, p.58a

Chapter 2 THE MIND-BODY CONNECTION

41. *Writings of the Ari*, Rabbi Isaac Luria, Vol. 12, Gate of the Holy Spirit, Research Centre of Kabbalah, ed., p.39
42. *Kabbalah for the Layman*, Vol. II, Berg, pp.162-164
43. *Kabbalah for the Layman*, Vol. I, Berg
44. *Zohar* III, p.58a
45. *The Mysterious Universe*, Sir James Jeans, AMS Press Reprint of 1933 Edition, p. 137

46. *Time Zones*, Part III, Berg
47. *Kabbalah for the Layman*, Vol. II, Berg, pp.132-135
48. *Kabbalah for the Layman*, Vol. II pp.173-178
49. *Zohar* Tazria 27:144-147, 28: 150-153
50. Leviticus 14:34
51. Rashi, Exodus, Ch. 35
52. Job 5:24
53. Kings II, 17:26
54. Leviticus 14:45
55. Exodus 21:1
56. *Zohar*, Mishpatim 1:1, 2:2
57. *Kabbalah for the Layman*, Vol. I, Berg, pp.86-88
58. *Zohar* II, p.7a-8a
59. Rosh Hashanah Lecture, Berg
60. *Writings of the Ari*, Rabbi Isaac Luria, Gate of Reincarnations, Vol.13, 1989 ed. Research Centre of Kabbalah
61. *Zohar Chadash*, Vol.20, 1988 ed., p.70, column 4
62. Jeremiah, 31:33
63. *Zohar* III, p.124b
64. *Zohar*, Vayikra 59:387-388
65. Isaiah 11:9
66. *Zohar*, Emor 24:129
67. Genesis 6
68. Genesis 6:6
69. Genesis 11
70. Genesis 11:5
71. *Zohar* I, p.75a; *Midrash Raba* Genesis, 38
72. Genesis 19
73 Genesis 13:13
74. *Writings of the Ari*, Rabbi Isaac Luria, Gate of Verses, Vol.8, 1989 ed., pp.101-103, Research Centre of Kabbalah

75. *Babylonian Talmud*, Tractate Gitten, 47a
76. *Kabbalah for the Layman*, Vol. III, Berg, pp.74-76

CHAPTER 3 HUMAN MYOPIA

77. *Zohar*, Shemot 15:96-112, 136-145
78. Isaiah 2:19
79. Isaiah 26:13
80. *Power of the Aleph Beth*, Vol. I, Berg, Ch.2
81. *Zohar*, Vaera 15:117-121
82. Exodus 7:12
83. *Kabbalah for the Layman*, Vol. I, Berg, pp.101-103
84. *Power of the Aleph Beth*, Vol. II, Berg, pp.33-35
85. *Zohar* II, p.26b
86. Genesis 2:9
87. Genesis 3:6,11
88. *Kabbalah Connection*, Berg, pp.117-119
89. Genesis 3:7
90. *Zohar*, Vaera 15:128-129
91. Exodus 7:19
92. Exodus 7:20
93. Isaiah 34:6
94. *Zohar*, Vaera 15:130-131
95. *Kabbalah Connection*, Berg, pp.96-100
96. *Power of the Aleph Beth*, Vol. I, pp.102-103
97. *Zohar* II, p.76
98. *Kabbalah Connection*, Berg, pp.142-146
99. Isaiah 34:6
100. *Zohar* II, p.28b
101. Daniel 12:10

102. *Power of the Aleph Beth*, Vol. I, Berg, pp.158-160

103. Daniel 12:3

104. *Zohar* III, p.124b

105. Esther 2:1-21

106. *Zohar*, Emor 24:129

107. *Kabbalah Connection*, Berg, pp.117-119

108. Isaiah 65:22

109. Exodus 32:4

110. *Zohar*, Beresheet A 52:482

CHAPTER 4 FISSION OR FUSION

111. *Kabbalah for the Layman*, Berg, Vol. II, pp.154-159

112. *Original Papers in Quantum Physics*, Max Planck, London, Taylor and Francis, 1972

113. *Zohar* III, p.99b

114. *Power of the Aleph Beth*, Vol. I, Berg, pp.107-108

115. *Power of the Aleph Beth*, Vol. II, pp.44-47

116. *Zohar*, Behar 8:52-57

117. Leviticus 25:20

118. Psalms 37:3

119. *Astrology: Star Connection*, Berg, pp.113-114

120. Exodus 16:29

121. *Zohar*, Vayera 32:460

122. Exodus 7:1-3

123. *Kabbalah for the Layman*, Vol. II, Berg, pp.119-120

124. *Zohar* I, p.193b

125. *Kabbalah for the Layman*, Vol. I, Berg, pp.78-80

126. *Zohar*, Parashat Pinchas, Vol. I, pp. xl-xliii

CHAPTER 5 STRESS

127. *Time Zones*, Berg, p.40
128. Genesis 4:1
129. *Zohar*, Beresheet B 60
130. *Astrology: Star Connection*, Berg, pp.148, 152
131. *Power of the Aleph Beth*, Vol. I, Berg, pp.216-218
132. *Power of the Aleph Beth*, Vol. II, Berg, pp.164-166
133. *Zohar*, Shemot 15:96-97
134. Zechariah 13:9
135. *Zohar*, Idra Raba 9:65
136. Jeremiah 31:33
137. *Zohar*, Idra Raba 9:65
138. Isaiah 11:9
139. *Kabbalah for the Layman*, Vol. I, Berg, pp.78-80
140. *Kabbalah for the Layman*, Vol. III, Berg, pp.141-144
141. *Kabbalah for the Layman*, Vol. III, Berg, pp.175-178
142. *Kabbalah for the Layman*, Vol. I, Berg, pp.85-86
143. *Kabbalah Connection*, Berg, pp.96-98

CHAPTER 6 THE OUTER AND INNER WORLD OF MAN

144. *Kabbalah Connection*, Berg, p.77
145. *Astrology: Star Connection*, Berg, pp.19-20
146. *Zohar*, Lech Lecha 31:333
147. Psalms 19:2-3
148. *Kabbalah for the Layman*, Vol. I, Berg, pp.102-108
149. *Power of the Aleph Beth*, Vol. I, Berg, pp.71-72
150. *Sefer Yetzirah* (*Book of Formation*), Yeshivat Kol Yehuda Press, Jerusalem, 1990

151. *Wheels of a Soul*, Berg, pp.110-112
152. *Zohar*, Tazria 33:169-175
153. Jeremiah 22:13
154. *Kabbalah for the Layman*, Vol. III, pp.148-149
155. *Wheels of a Soul*, Berg, pp.153-154
156. *Kabbalah for the Layman*, Vol. II, Berg, pp.149-151
157. *Kabbalah for the Layman*, Vol. III, Berg, pp.162-164
158. *Kabbalah for the Layman*, Vol. II, Berg, pp.70-74
159. *Ten Luminous Emanations*, Vol. III, Rabbi Yehuda Ashlag, Heb. Ed., Research Centre of Kabbalah
160. *Kabbalah for the Layman*, Vol. II, Berg, pp.126-127
161. *Writings of the Ari*, Gate of Introductions, Vol.6, Research Centre of Kabbalah, p.211
162. *Writings of the Ari*, Tree of Life, Rabbi Isaac Luria, Gate 42, Chapter 1, Research Centre of Kabbalah
163. *Zohar*, Vayak'hel 25:369-373
164. Lamentations 3:23
165. *Kabbalah for the Layman*, Vol. III, Berg, pp.159-161
166. *Kabbalah for the Layman*, Vol. II, Berg, pp.38-39
167. *Power of the Aleph Beth*, Vol. II, Berg, p.91
168. *Zohar*, Vayishlach 1:1-4
169. Genesis 32:4
170. Psalms 91:11
171. Genesis 4:7
172. Psalms 51:5
173. *Zohar*, Vayashev 1:1-3
174. Genesis 37:1
175. Psalms 34:20
176. Genesis 4:7
177. Ecclesiastes 4:13
178. Ecclesiastes 2:14

179. *Zohar*, Parashat Pinchas, Vol. II, Berg
180. *Wheels of a Soul*, Berg, pp, 58, 59
181. *Power of the Aleph Beth*, Vol. I, pp.146-148.
182. *Kabbalah for the Layman*, Vol. II, pp.1l9-123
183. *Kabbalah for the Layman*, p.122
184. *Kabbalah for the Layman*, Vol. III, Berg, pp.142-144
185. *Kabbalah for the Layman*, pp.166-167
186. *Power of the Aleph Beth*, Vol. II, Berg, pp.160-161
187. *Zohar* I, p.78a
188. *Power of the Aleph Beth*, Vol. I, Berg, pp.102-103
189. *Kabbalah for the Layman*, Vol. I, Berg, pp.79-90
190. *Power of the Aleph Beth*, Vol. I, Berg, p.l05
191. Psalms 145:18
192. *Zohar*, Ha'azinu 51:210
193. Genesis 4:1
194. *Zohar* I, p.54a
195. Micah 7:20
196. Psalms 145:18
197. *Kabbalah for the Layman*, Vol. II, Berg, pp.106-108
198. *Zohar*, Ha'azinu 51:210

CHAPTER 7 IMMUNE-VULNERABLE

199. Genesis 32:2
200. Genesis, 32:25
201. Genesis, 32.11
202. *Zohar*, Vayishlach 1:5, 10, 12
203. Proverbs 12:9
204. Genesis 4:7
205. *Zohar* Vayishlach 1:16, 17, 19

206. Genesis, 32:24-25
207. *Power of the Aleph Beth*, Vol. I, Berg, p.59.
208. *Wheels of a Soul*, Berg, p.127
209. *Power of the Aleph Beth*, Vol. I, Berg, pp.67-69
210. *Zohar*, Pinchas 29:167
211. *Kabbalah Connection*, Berg, p.101
212. *Zohar*, Pinchas 68:410
213. *Zohar*, Shoftim 2:2
214. *Zohar* II, p.265a
215. Proverbs 30:21-23
216. *Zohar*, Shemot 41:306-308
217. *Zohar*, Vayetze 15:111-112
218. *Astrology: Star Connection*, Berg, p.109
219. Genesis 6:12
220. *Zohar*, Noah 25:192-193
221. *Kabbalah for the Layman*, Berg, Vol. II, pp.l06-107
222. *Zohar*, Lech Lecha 2:4
223. Genesis 6:9-22

CHAPTER 8 TIME TRAVEL

224. *Astrology: Star Connection*, Berg, pp.16-21
225. *Astrology: Star Connection*, Berg, pp.166-169
226. *Kabbalah for the Layman*, Vols. I, II, III; *Power of the Aleph Beth*, Vols. I, II, Berg
227. *Kabbalah for the Layman*, Vol. I, Berg, p.149
228. *Zohar* II, p.81b
229. *Zohar* III, p.125a
230. *Zohar* II, p.81a
231. Exodus 32

232. *Time Zones*, Berg, p.89
233. *Kabbalah for the Layman*, Vol. I, p.36; Zohar III, pp.194a-b
234. *Astrology: Kabbalah Connection*, Berg, p.34
235. *Zohar*, Pinchas 77:480
236. *Zohar*, Vayak'hel 17:223
237. *Power of the Aleph Beth*, Vol. I, Berg, p.96; *Star Connection*, Berg, p.72
238. *Kabbalah for the Layman*, Vol. III, Berg, pp.115-118
239. Genesis 2:9
240. *Power of the Aleph Beth*, Vol. I, Berg, p.96; *Star Connection*, Berg, p.72
241. Genesis 28:11-13
242. *Kabbalah for the Layman*, Vol. III, Berg, pp.179-188
243. *Astrology: Star Connection*, Berg, pp.21-22
244. *Power of the Aleph Beth*, Vol. I, Berg, p.110
245. *Zohar* II, p.206a

GLOSSARY

ADAM – The *Sefirot* represented as a man: *Keter* as the brain; *Chochmah* the eyes; *Binah* the ears; *Zeir Anpin* the nose; *Malchut* as the mouth.

ADAM AND EVE – From the kabbalistic point of view, Adam and Eve form one undifferentiated soul. After the fall they became two parts of one soul—soul mates. Adam representing the male principle of drawing energy for sharing; Eve representing the female principle of receiving and revealing.

ADAR – The twelfth month of the Kabbalistic calendar lunar year, sixth month from *Rosh Hashanah*. Approximating to February/March. Its zodiac sign is Pisces.

AGE OF AQUARIUS – The Age of Messiah, ushered in by the *Writings of the Ari* Z"L (Rabbi Isaac Luria) forty years after the expulsion from Spain. From that time onward, many of the limitations and prohibitions surrounding kabbalistic study were completely removed.

AGGADAH – Name of those sections of *Talmud* and *Midrash* containing homiletic expositions of the Bible.

ALEF BET – The metaphysical DNA of all Creation that channels Light into our world through 22 letters that manifest in the Hebrew writing system.

ALL-EMBRACING UNIFIED WHOLE – The Lord.

ANGELS – Manifested Supernal energy intelligences, beings of Light dedicated to specific purposes who are not subject to Free-Will.

ASHKENAZI – Referring to West, East or Central European Jewry, as contrasted with Sephardim. (See also, "PARDES")

ASTRAL TRAVEL – The non-corporeal mode of traveling great distances that transcends time, space, and motion.

BAR/BAT MITZVAH – The time when the aspect of imparting is awakened in the soul—age thirteen in the male and twelve in the female.

BERICH SHEMEI – A prayer in Aramaic said before taking out the Torah from the Ark. The power of the prayer is the transcending of time, space, and motion; a spiritual time tunnel that returns our consciousness back to the giving of the Torah.

BERACH – Benediction or blessing, the metaphysical connection to the internal energy-intelligence of things.

BET DIN – Rabbinical court of law.

BODY ENERGY-INTELLIGENCE – The energy-intelligence of the Desire to Receive for the Self Alone.

BOOK OF DANIEL – The Prophet Daniel lived in Persia at the time of Nebuchadnezzar. The *Book of Daniel* contains coded knowledge referring to the Age of Aquarius and the letters of the Hebrew Alef Bet; the wisdom of the *Book of Daniel* is sealed until the end of days.

BOOK OF ESTHER – Or *Megillat Esther* (which means the revealment of the concealed), the festival scroll of *Purim*, telling the story of the salvation of the Jews of Persia. It speaks in length about the giving of gifts and charity; it unveils many great secrets and contains coded information on overcoming all evil.

BOOK OF SPLENDOR* or *ZOHAR – Of the *Zohar* it is written: In your compendium, Rabbi Shimon Bar Yochai, the *Zohar*, shall Israel and the world in the future taste from the Tree of Life, which is the *Book of Splendor*. And the world shall go forth from its exile with mercy. (Zohar, Naso 6:90)

BOTZ – World of Mud. This physical mundane plane that maintains its grip as our daily reality.

BRAIN – The physical corporeal vessel that permits the manifestation of the mind.

BREAD OF SHAME – The shame felt when receiving something for nothing. The whole purpose of life in this world is to remove the Bread of Shame.

CABLES – Various means for the transference of positive metaphysical energies to humanity (such as, prayer, meditation, *Shabbat*, cosmic time zones, etc.).

CANCER – Sign of the zodiac corresponding to the Hebrew month of *Tammuz* in which the dreaded disease of the same name may begin due to vulnerability caused by a crack in the security shield of positivity. A time in which extra care must be taken to avoid falling into arguments and other negative activities.

CAUSE – That which brings about the revelation of a level.

CENTRAL COLUMN – The connecting, balancing link between the Right and Left Columns of positive and negative energy, male and female. Aspect of Restriction.

CHARIOTS – Entities that embody both metaphysical and physical levels of energy-intelligence.

CHESED – Mercy, Loving Kindness. First of the Seven Lower *Sefirot*. The Right Column Abraham the Patriarch is the chariot of *Chesed*.

CHOCHMAH – Wisdom. The second *Sefira* following *Keter* and is the architectural bottled up energy of total creation.

CIRCULAR CONCEPT – The balance between Left and Right, negative and positive, brought about by use of Restriction. Central Column.

COLUMNS – (Right, Left and Central) Macrocosmic pipes or lines of energy corresponding to positive, negative, and balancing energies, similar to the proton, electron, and neutron in the microcosmic atom.

CONSCIOUSNESS – Levels of awareness. As the soul sheds the veils of negativity caused by the Desire to Receive for the Self Alone, higher levels of understanding and awareness are made manifest.

CONSCIOUSNESS, ALTERED STATE OF – A state of conscious awareness that transcends the five physical senses. An enhanced and elevated level of consciousness brought about by developing the Desire to Receive for the Sake of Sharing and complete circuitry and connection to the Light.

CONSCIOUSNESS, BODY – The Desire to Receive for the Self Alone.

CONSCIOUSNESS, COSMIC – The highest state of awareness where the universe is conceived of as one all embracing unity, in which all souls are recognized as one indivisible, interrelated, and interdependent whole beyond the confines of time, space, and motion. Quantum consciousness, where past, present, and future are unified in the now.

CONSCIOUSNESS, SOUL – The Desire to Receive for the Sake of Sharing.

CORRECTION – The task of bringing cosmic and individual harmony to the universe in a state of perfection.

COSMIC DANGER ZONES – Cyclically occurring zones of time, which manifest strong negative influences and can either be overcome or at least mitigated by the knowledge of Kabbalah and the use of Restriction.

COSMIC INFLUENCES – Just as the moon influences the tides of all bodies of water on Earth on a physical level, and emotional states on a more subtle level, so do the myriad cosmic influences combine and interweave, metaphysically as well as physically, to shape the destiny of humanity and the universe. Man, with the proper knowledge of kabbalistic tools, has the ability to take control over these and make manifest a higher parallel reality of harmony and peace.

CREATOR – The source of all positive energy to the total exclusion of any negative energy.

DA'AT – Knowledge.

DALET – "Poor," also the fourth letter of the Hebrew Alphabet, symbolizing the Earth with the connection of either the *Shechinah* or *Zeir Anpin*.

DAVID, KING – Chariot of Malchut. Second King of Israel and tribe of Judah, succeeding Saul. Author of many Psalms. King David's son by Batsheva, King Solomon, constructed the First Temple.

DELUGE – The great flood described in the biblical account of Noah, Genesis 7.

DESIRE TO RECEIVE FOR THE SELF ALONE – Negativity. The aspect of drawing or taking. In our universe all is made up of the Desire to Receive. On the physical level, a Desire to Receive for the Self Alone is characterized by selfishness, egotism, materialism in man. Our purpose is to transform this selfish desire into a Desire to Receive for the Sake of Sharing, a balance and harmony between receiving and imparting, permitting the individual to draw into himself the positive Light of the Creator.

DESIRE TO RECEIVE FOR THE SAKE OF SHARING – Balance. The aspect of receiving in order to share, giving as opposed to the Desire to Receive for the Self Alone.

DNA – Deoxyribonucleic acid. Bound in double helical chains forming the basic material in the chromosomes of the cell nucleus, it contains the genetic code and transmits the hereditary pattern.

DOR DE'AH – The Generation of Knowledge, originally the Generation of the Flood, which reincarnated at the time of the Tower of Babel, and again during the Exodus, and now during the Age of Aquarius.

DEVEKUT – "Cleaving." Fulfillment of the circular concept whereby union is brought about between the Light of the Creator and man.

EGO – The individual as self-aware, self-centered. From the kabbalistic point of view, ego is the manifestation of the Desire to

Receive for the Self Alone. Ego is the underlying factor for the limited expression of our five percent consciousness. Our ego convinces us that all our decisions and activities are the direct result of our conscious mind and thought.

EMPTY SPACE – Vacuum, non revealment of the Lightforce. This void represents the energy intelligence of vulnerability.

ENCIRCLING LIGHT CONSCIOUSNESS – Superconscious Encircling Light takes off where Inner Light consciousness ends. The all pervading consciousness of the cosmos, where information of past, present, and future meet as one unified whole. It extends beyond the inner light consciousness of humankind. It is precisely Encircling Light consciousness that we find most in our lives.

ENDLESS, THE – The Infinite All Embracing Unity.

ENERGY TRANSFER SYSTEM – Transfer systems as prescribed within the Wisdom of the Kabbalah, created and written on special parchment by qualified scribes, to provide cosmic consciousness and pure awareness to those seeking a higher level of cosmic intelligence through the power of *Tefillin*, *Mezuzot*, *Megillot*, and *Sefer Torah*.

EVIL EYE – There are some men specially fitted for the transmission of blessings, as for instance a person of a "good eye." There are others who are specially fitted for the transmission of negativity and curses. "On whatever their eyes fall their curses are confirmed.... Hence, a man should turn aside a hundred times in order to avoid a man with an evil eye.

FISSION – The process of fragmentation and disunity, i.e. splitting the nucleus of an atoms. (See Fusion)

FISSION CONSCIOUSNESS – The level of consciousness in which separation, fragmentation, and disunity reign. As opposed to Fusion Consciousness in which a quantum, holistic awareness of Unity prevail.

FIVE SENSES – The senses of our body-consciousness: sight, hearing, smell, taste, touch.

FRAGMENTATION – Disunity and disruption brought about by the divisive and destructive manifestation of the Desire to Receive for the Self Alone.

FREE-WILL – The ability to choose between manifesting the Desire to Receive for the Sake of Sharing or the Desire to Receive for the Self Alone.

FUSION – The union of different things by or as if by melting, blending, coalition. (See Fission)

FUSION CONSCIOUSNESS – The consciousness of uniting, bringing together parts to form a whole. From a kabbalistic point of view, this is the awareness of quantum reality.

GEMAR HATIKKUN – The Final Redemption of Israel, the ultimate peace and harmony in the world. (See Correction)

GOLDEN CALF – A calf of gold worshiped by the erev rav (mixed multitude) while Moses was at Mount Sinai.

GUT (Grand Unified Theory) – During the 1970s, fundamental physics set out to unify the strange and complex world about us into a single conceptual framework. Fresh discoveries have opened the way to a radical new concept of a unified universe.

GEVURAH – The *Sefira* of Judgment, power, might. The second of the Seven Lower *Sefirot*. Left Column, Isaac is the chariot of *Gevurah*.

HEISENBERG'S UNCERTAINTY PRINCIPLE – It is the concept, according to quantum mechanics, that it is impossible to measure two related quantities, simultaneously and exactly. From a kabbalistic point of view, the basis of this dilemma is a fundamental rule. The conclusion brought to the forefront is the notion that things can be everywhere at once; no space, as the *Zohar* determined.

HEI – The second and the fourth letter of the sacred Tetragrammaton, the first *Hei* representing the *Sefira Binah* and the second, the *Sefira Malchut*.

HOD – Splendor. Fifth of the Seven Lower *Sefirot*, the Left Column. Aaron the High Priest is the chariot of Hod.

INTELLIGENCE – Reflection on the ways of cause and effect in order to clarify the final result.

JERUSALEM – Holy inasmuch as it portrays a constant flow of internal energy. The energy center of the world.

KABBALAH – The inner soul of the Torah. From the Hebrew *lekabel*, meaning "to receive."

KABBALISTIC MEDITATION – Special techniques of meditation, which are fully described in the *Writings of the Ari* Z"L.

KAVANAH – The need to center one's inner world with the attention appropriate to the situation or connection.

KETER – Crown. The link between the Lightforce of the Creator and the brain is *Keter*, the seed of all physical manifestation and activity. First of the Upper Three *Sefirot* of the Ten *Sefirot*.

KLIPOT – (Sing. *klipa*) Shells, evil husks created by man's negative deeds that cover and limit man in his spiritual development. The barriers between man and the Lightforce of the Creator.

LEFT COLUMN – The channel through which all metaphysical energies are drawn. (See Desire to Receive).

LIGHT BARRIER – The Desire to Receive for the Self Alone that impedes the revealment of the Light.

LIGHT – The source and force of all energy, mental and physical with an intrinsic characteristic of sharing.

LIGHTFORCE, THE – Lord, the Light; the All–Embracing Unity. (See Light)

LURIANIC KABBALAH – The system of kabbalistic inquiry and practice as established by Rabbi Isaac Luria. Emphasizes the more active side of prayer, and deals specifically with the sparks of Light that elevate in prayer. In kabbalistic literature prayer is like an arrow shot upward by the reciter with the bow of *kavanah*.

MALCHUT – Kingdom. The tenth and final *Sefira* from *Keter*. The *Sefira* in which the greatest Desire to Receive is manifest and in which all correction takes place. The physical world.

NESHAMA – Third of the five levels of the soul. Correlated with the *Sefira* of *Binah.*

NETZACH – Victory. Fourth of the Seven Lower *Sefirot.* Synonymous with the right column. Moses, symbolized as its chariot.

NON CORPOREAL BEINGS – Thought conscious extraterrestrial entities without the physical limitations of time, space, and motion.

OR EIN SOF – The Light of the Infinite (Endless) from which sprang all future emanations. The primal Light in which the souls of man were in perfect harmony with the Creator. A complete balance between the endless imparting of the Creator and the endless receiving of his creations—the souls of man. That of which nothing can be understood and which yet must be postulated.

PARALLEL UNIVERSES – The realms of the Tree of Life Reality and the illusionary Tree of Knowledge Reality as outlined in Genesis.

PARDES – Biblical interpretation of verses or words of the Bible. Consists of four Hebrew letters: Pei for *peshat* or literal), Resh for *remez* or allegorical, Dalet for *derash* or political, Sin for *sod* or esoteric/Kabbalah. The *Zohar* says: "Four persons entered the *pardes* (orchard) concerning the nature and process of Creation: Ben Azzai, Ben Zoma, Acher (meaning other and the surname given to Elisha Ben Abuyah) and Rabbi Akiva. Ben Azzai, Ben Zoma and Acher entered the domains of *peshat, remez* and *derash* interpretations of the Torah. Only Rabbi Akiva entered the domain of *sod,* and he alone survived. The others who entered PeReD, which the world of separation did not survive. It is through the addition of the letter *Sin* (Kabbalah) that the word PeReD (separation), is changed to PaRDeS, the world of unity. These four letters also form the word SePheRaD. So one studying the Kabbalah is termed Sepharadi. The Hebrew word Sepharad (Jews generally considered as originally from Spain), is one of the most misinterpreted and misunderstood words ever to have emerged with Judaism. Consequently, because of the *tikkun* process, one may be incarnated as an Ashkenazi Jew or one who studies PeReD, but if his studies also include the *Sod* (Kabbalah), then he is in essence a Sephardic Jew. Contrarily, one incarnated as a Sephardi and neglects or is even opposed to the study of Kabbalah, is considered an Ashkenazi Jew. (*Zohar* 1, p. 26b, 27a).

PLACEBO EFFECT – The effect of a placebo, a harmless, un-medicated preparation, given to a patient merely to humor him, or used as a control in testing the efficacy of another medicated substance can demonstrate the power of the human mind in healing under the given positive suggestion. The placebo effect demonstrates the psychosomatic quality of disease.

QUANTUM – In the kabbalistic sense of the word: The substance of quantum is "Love Thy Neighbor." When this is achieved by mankind, the entire universe, both the seen and the unseen, will be revealed as it actually is, a single unified whole. Our universe is perceived as fragmented only because mankind is fragmented.

RADIOACTIVE WASTE – Hazardous materials resulting from man's tampering with the intrinsic balance of the universe.

REINCARNATION – From a kabbalistic point of view it is the movement and stages the soul journeys to achieve its *tikkun* (correction).

R.E.M. – Rapid eye movement. During sleep, periods of rapid eye movement indicate the occurrence of dreams.

RESTRICTION – The Central Column energy intelligence force that establishes and maintains balance in the universe.

ROBOTIC CONSCIOUSNESS – When celestial influences govern the daily activities of man without intervention of his/her intrinsic ability to exercise free-will.

RIGHT COLUMN – Chesed. Column that draws the energy of imparting the positive force.

RUACH – Second lower level of soul consciousness. Prior to the sin of Adam, which was the negation of the Lightforce, the entire universe existed and remained connected at the level of *Ruach*, unfettered by the claims of space and time, unshadowed by entropy and death. Associated with the *Sefira* of *Zeir Anpin*.

SATAN – The personification of the Desire to Receive for the Self Alone.

SCAN – From a kabbalistic point of view, the human eye is the window of the soul and as such is a powerful tool for the transmission and reception of the Light channeled by the letters and words of the *Zohar*. The connection is established at the metaphysical level of our being and radiates into our physical plane of existence. Hebrew reads Right to Left.

SECURITY SHIELD – When the Shield of David is activated a protective film of the Lightforce surrounds the individual thereby preventing the invasion of the Dark Lord and its devastating fleet of misery and illness.

SEFER YETZIRA – *Book of Formation*. First known kabbalistic work containing in concise, highly esoteric language, the entire teachings of Kabbalah. The first written work of the Kabbalah. Attributed to the patriarch Abraham.

SEFIRA (pl. *Sefirot*) – A kabbalistic term denoting the ten spheres or metaphysical channels or vessels through which the Lightforce of the Creator, emanates and manifests itself, and is emanated to man.

SEPHARADI – (pl. *Sepharadim*), see PARDES.

SHECHINAH – Cosmic realm to which an individual may connect and acquire cosmic consciousness, manifested in the dimension of *Malchut*.

SITREI TORAH – The deepest hidden teachings of the Torah received only through Divine revelation.

SLEEP – From a kabbalistic point of view sleep permits the soul to temporarily extricate itself from the limitations and uncertainties of the physical body consciousness.

SOUL – The Light clothed in the Vessel of Intelligence.

SPACE-TIME – Where time is now addressed as a gap or empty space.

SPEED OF LIGHT – 186,000 miles per second. From the kabbalistic point of view, light does not travel but is ever present albeit in a state of concealment awaiting revealment. The kabbalist, therefore, speaks of the "speed" of the revealment of light.

TA'AMEI TORAH – The reasons (tastes of Torah) through which one reaches the true inner meanings of Torah and thereby elevates oneself to the highest degrees of spirituality.

TALMUD – The written form of the oral law. The main work of Judaic studies. A compilation of *Mishna, Tosefot, Gemara*.

TAMMUZ – Fourth month of the Jewish calendar lunar year, tenth from *Rosh Hashanah*, approximating to June/July. Its zodiac sign is Cancer.

TEMPLE – A physical structure over the energy center of the universe acting as the receptacle or receiving station for the Lightforce of the Creator.

TETRAGRAMMATON – The sacred Name composed of the four Hebrew letters, *Yod, Hei, Vav,* and *Hei*.

THE BOOK OF FORMATION – See *Sefer Yetzirah*.

THE POWER OF ONE – The All Embracing Unified Whole, the Lightforce.

THOUGHT CONSCIOUSNESS – The only true reality that must be considered in a frame of energy intelligence.

TIKKUN – The process of correction made by the soul.

TREE OF LIFE – The point from which the life energy force remains as an all embracing unified whole without the trappings of chaos and uncertainty.

TREE OF KNOWLEDGE GOOD AND EVIL – Realm of our illusionary reality. Here randomness, uncertainty, chaos, rot, disorder, illness and misfortune make their presence felt.

TESHUVA – No individual can ever achieve a completed phase of *teshuva*, (a back to the future concept) where the individual comes into total control of his fate and destiny, unless he becomes knowledgeable of the unconscious root processes of the soul along with the knowledge of former lifetimes.

TZADIK – Righteous. Associated with the *Sefira* of *Yesod* and the Covenant.

TZIMTZUM – The original Restriction.

UNCONSCIOUS PROCESSES – Mental processes that one is unable to bring into one's conscious mind.

VULNERABILITY – Openness to attack, injury.

WHOLISTIC – From a kabbalistic point of view pertaining to the complete picture, the quantum picture, the complete circuitry. Concerned with whole or integrated systems rather than their parts. As opposed to atomistic.

WISDOM – (*Chochmah*) The second *Sefira* of the Upper Three *Sefirot* and the first of the Four Phases. Knowledge of the final ends of all aspects of reality.

YESOD – Sixth of the Seven Lower *Sefirot* of which Joseph is the chariot. The *Sefira* through which is emanated all light to our world.

YOD – Smallest and yet most powerful letter of the *Alef Bet*. The first and initial letter of the Tetragrammaton.

ZOHAR – The foundational source of the Kabbalah. Written by Rabbi Shimon bar Yochai while hiding from the Romans in a cave in Peki'in for 13 years. Later brought to light by Rabbi Moses de Leon in Spain.

About the Centres

Kabbalah is the deepest and most hidden meaning of the Torah or Bible. Through the ultimate knowledge and mystical practices of Kabbalah, one can reach the highest spiritual levels attainable. Although many people rely on belief, faith, and dogmas in pursuing the meaning of life, Kabbalists seek a spiritual connection with the Creator and the forces of the Creator, so that the strange becomes familiar, and faith becomes knowledge.

Throughout history, those who knew and practiced the Kabbalah were extremely careful in their dissemination of the knowledge because they knew the masses of mankind had not yet prepared for the ultimate truth of existence. Today, kabbalists know that it is not only proper but necessary to make the Kabbalah available to all who seek it.

The Kabbalah Centre is an independent, non-profit institute founded in Israel in 1922. The Centre provides research, information, and assistance to those who seek the insights of Kabbalah. The Centre offers public lectures, classes, seminars, and excursions to mystical sites at branches in Israel and in the United States. Branches have been opened in Mexico, Montreal, Toronto, Paris, Hong Kong and Taiwan.

Our courses and materials deal with the *Zoharic* understanding of each weekly portion of the Torah. Every facet of life is covered and other dimensions, hithertofore unknown, provide a deeper connection to a superior reality. Three important beginner courses cover such aspects as: Time, Space and Motion; Reincarnation, Marriage, Divorce; Kabbalistic Meditation; Limitation of the Five Senses; Illusion-Reality; Four Phases; Male and Female, Death, Sleep, Dreams; Food; and Shabbat.

Thousands of people have benefited from the Centre's activities, and the Centre's publishing of kabbalistic material continues to be the most comprehensive of its kind in the world, including translations in

English, Hebrew, Russian, German, Portuguese, French, Spanish, and Farsi (Persian).

Kabbalah can provide one with the true meaning of their being and the knowledge necessary for their ultimate benefit. It can show one spirituality that is beyond belief. The Kabbalah Centre will continue to make available the Kabbalah to all those who seek it.

About the *Zohar*

The *Zohar*, the basic source of the Kabbalah, was authored, two thousand years ago, by Rabbi Shimon bar Yochai while hiding from the Romans in a cave in Peki'in for 13 years. It was later brought to light by Rabbi Moses de Leon in Spain, and further revealed through the Safed Kabbalists and the Lurianic system of Kabbalah.

The programs of The Kabbalah Centre have been established to provide opportunities for learning, teaching, research, and demonstration of specialized knowledge drawn from the ageless wisdom of the *Zohar* and the Jewish sages. Long kept from the masses, today this knowledge of the *Zohar* and Kabbalah should be shared by all who seek to understand the deeper meaning of this spiritual heritage, and a deeper and more profound meaning of life. Modern science is only beginning to discover what our sages veiled in symbolism. This knowledge is of a very practical nature and can be applied daily for the betterment of our lives and of humankind.

Darkness cannot prevail in the presence of Light. Even a darkened room must respond to the lighting of a candle. As we share this moment together we are beginning to witness, and indeed some of us are already participating in, a people's revolution of enlightenment. The darkened clouds of strife and conflict will make their presence felt only as long as the Eternal Light remains concealed.

The *Zohar* now remains an instrument to infuse the cosmos with the revealed Lightforce of the Creator. The *Zohar* is not a book about religion. Rather, the *Zohar* is concerned with the relationship between the unseen forces of the cosmos, the Lightforce, and the impact on humanity.

The *Zohar* promises that with the ushering in of the Age of Aquarius, the cosmos will become readily accessible to human understanding. It states that in the days of the Messiah "there will no longer be the necessity for one to request of his neighbor, teach me wisdom." (*Zohar Naso*, 9:65) "One day, they will no longer teach every man his neighbor and every man his brother, saying know the Lord. For they shall all know Me, from the youngest to the oldest of them. (*Jeremiah* 31:34)

We can regain dominion of our lives and environment. To achieve this objective, the *Zohar* provides us with an opportunity to transcend the crushing weight of universal negativity.

The daily perusing of the *Zohar*, without any attempt at translation or understanding will fill our consciousness with the Light, improving our well-being, and influencing all in our environment toward positive attitudes. Even the scanning of the *Zohar* by those unfamiliar with the Hebrew *Alef Bet* will accomplish the same result.

The connection that we establish through scanning the *Zohar* is one of unity with the Light of the Creator. The letters, even if we do not consciously know Hebrew or Aramaic, are the channels through which the connection is made and can be likened to dialing a telephone number or typing in codes to run a computer program. The connection is established at the metaphysical level of our being and radiates into our physical plane of existence. But first there is the prerequisite of metaphysical "fixing." We have to consciously, through positive thought and actions, permit the immense power of the *Zohar* to radiate love, harmony, and peace into our lives for us to share with all humanity and the universe.

As we enter the years ahead, the *Zohar* will continue to be a people's book, striking a sympathetic chord in the hearts and minds of those who long for peace, truth, and relief from suffering. In the face of crises and catastrophe, the *Zohar* has the ability to resolve agonizing human afflictions by restoring each individual's relationship with the Lightforce of the Creator.

Kabbalah Centre Books

72 Names of God, The: Technology for the Soul
72 Names of God for Kids, The: A Treasury of Timeless Wisdom
72 Names of God Meditation Book, The
And You Shall Choose Life: An Essay on Kabbalah, the Purpose of Life, and Our True Spiritual Work
Angel Intelligence: How Your Consciousness Determines Which Angels Come Into Your Life
AstrologiK: Kabbalistic Astrology Guide for Children
Becoming Like God: Kabbalah and Our Ultimate Destiny
Beloved of My Soul: Letters of Our Master and Teacher Rav Yehuda Tzvi Brandwein to His Beloved Student Kabbalist Rav Berg
Consciousness and the Cosmos (Previously *Star Connection*)
Days of Connection: A Guide to Kabbalah's Holidays and New Moons
Days of Power Part 1
Days of Power Part 2
Dialing God: Daily Connection Book
Education of a Kabbalist
Energy of the Hebrew Letters, The (Previously Power of the *Aleph Beth Vols. 1 and 2*)
Finding the Light Through the Darkness: Inspirational Lessons Rooted in the Bible and the Zohar
God Wears Lipstick: Kabbalah for Women
Holy Grail, The: A Manifesto on the Zohar
If You Don't Like Your Life, Change It!: Using Kabbalah to Rewrite the Movie of Your Life
Immortality: The Inevitability Of Eternal Life
Kabbalah Connection, The: Preparing the Soul For Pesach
Kabbalah for the Layman
Kabbalah Method, The: The Bridge Between Science and the Soul, Physics and Fulfillment, Quantum and the Creator
Kabbalah On The Sabbath: Elevating Our Soul to the Light
Kabbalah: The Power To Change Everything

Kabbalistic Astrology: And the Meaning of Our Lives
Kabbalistic Bible: Genesis
Kabbalistic Bible: Exodus
Kabbalistic Bible: Leviticus
Kabbalistic Bible: Numbers
Kabbalistic Bible: Deuteronomy
Life Rules: How Kabbalah Can Turn Your Life From a Problem into a Solution
Living Kabbalah
Light of Wisdom: On Wisdom, Life, and Eternity
Miracles, Mysteries, and Prayer Volume 1
Miracles, Mysteries, and Prayer Volume 2
Nano: Technology of Mind Over Matter
Navigating The Universe: A Roadmap for Understanding the Cosmic Influences that Shape our Lives (Previously *Time Zones*)
On World Peace: Two Essays by the Holy Kabbalist Rav Yehuda Ashlag
Prayer of the Kabbalist, The: The 42-Letter Name of God
Power of Kabbalah, The: 13 Principles to Overcome Challenges and Achieve Fulfillment
Rebooting: Defeating Depression With The Power of Kabbalah
Satan: An Autobiography
Secret, The: Unlocking the Source of Joy & Fulfillment
Secrets of the Bible: Teachings From Kabbalistic Masters
Secrets of The Zohar: Stories and Meditations to Awaken the Heart
Simple Light: Wisdom from a Woman's Heart
Shabbat Connections
Taming Chaos: Harnessing the Secret Codes of the Universe to Make Sense of Our Lives
Thought of Creation, The: On the Individual, Humanity, and The Ultimate Perfection
To Be Continued: Reincarnation & the Purpose of Our Lives
True Prosperity: How to Have Everything
Vokabbalahry: Words of Wisdom for Kids to Live By
Way Of The Kabbalist, The: A User's Guide To Technology For The Soul

Well of Life: Kabbalistic Wisdom From a Depth of Knowledge
Wheels of a Soul: Kabbalah and Reincarnation
Wisdom of Truth, The: 12 Essays by the Holy Kabbalist Rav
     Yehuda Ashlag
Zohar, The

BOOKS AVAILABLE AT
WWW. KABBALAH.COM/STORE
AND KABBALAH CENTRES AROUND THE WORLD

May the Light that this knowledge brings to the world,
guide you on your way back to the Light.

With all our love,
Marco, Ursula and Fabiola